629.222 F986a
Furman, Michael.
American auto legends :
31112016716884
WITHDRAWN
01142011

Do Not Use Book Return!
Return Inside
ROCKFORD PUBLIC LIBRARY
Rockford, Illinois
www.rockfordpubliclibrary.org
815-965-9511

DEMCO

AMERICAN
AUTO LEGENDS

Ford
MADE IN
U.S.A.

AMERICAN AUTO LEGENDS

Classics of Style and Design

Michael Furman

Text by Tracy Powell

ROCKFORD PUBLIC LIBRARY

First published 2010 by

Merrell Publishers Limited
81 Southwark Street
London SE1 0HX

merrellpublishers.com

All illustrations copyright © 2010 Michael Furman, with the exception of the following:
pages 8, 9, 10, 11, 12, 13, 17, 18, 19, 20, 21, 22, 24, 25 courtesy of Automobile Quarterly Photo and Research Archive

Text, design, and layout copyright © 2010 Merrell Publishers Limited

All rights reserved. No part of this publication may be reproduced, stored in a retrieval system, or transmitted, in any form or by any means, electronic, mechanical, photocopying, recording, or otherwise, without the prior written permission of the publisher.

A catalogue record for this book is available from the Library of Congress.

British Library Cataloguing-in-Publication Data:
Furman, Michael.
American auto legends : classics of style and design.
1. Automobiles – United States – Design and construction – History. 2. Automobiles – United States – Design and construction – History – Pictorial works.
I. Title II. Powell, Tracy.
629.2′22′0973-dc22

ISBN 978-1-8589-4516-3

Produced by Merrell Publishers Limited
Layout by Jonny Burch
Design concept by Matt Hervey
Project-managed by Mark Ralph
Proof-read by Richard Mason
Indexed by Vicki Robinson

Printed and bound in China

Jacket, front: Auburn 851 Speedster, see page 115
Jacket, back: Ford Mustang GT350H, see page 223
Page 2: Ford Model T Runabout, see page 47
Pages 28–29: Stutz Model M Speedster, see page 84
Pages 276–77: Corvette, see page 170

Acknowledgments

My deepest appreciation goes to the many collectors and museum curators who have made their wonderful cars available to me over the years. Special thanks go to the dedicated people at Merrell Publishers and my staff, including senior artist Dave Phillips, graphic specialist Mary Dunham, and project manager Phil Neff. It is a pleasure to work with such talented people in order to bring you these photographs.

Michael Furman

I wish to thank my wife, Cristine, whose patience and support is irreplaceable; and the publisher of *Automobile Quarterly*, Gerry Durnell, who gave me my first shot at writing and editing in the automotive realm a decade ago. The effort put into the research for this book is dedicated to the memory of Beverly Rae Kimes, to whom I owe much gratitude for the few occasions on which we shared the challenge of keeping the flame alive.

Tracy Powell

INTRODUCTION

From the boons of Henry Ford's assembly line and Dwight D. Eisenhower's Interstate Highway System to the banes of suburban sprawl and foreign-oil dependence, U.S. history is closely tied to the automobile. Thanks to enthusiasts living in the United States and abroad, automotive archeology and preservation have left us with gems of an American heritage that we recognize through such marques as Packard, Cadillac, Duesenberg, Locomobile, Studebaker, and many others.

The American automotive experience can be summed up in numbers. Unlike their European counterparts, American automakers proliferated in the dozens, then the hundreds, producing thousands of different models. Sources differ on the actual number of automobile manufacturers originating in the United States, ranging from several hundred to a few thousand. One thing is sure: of the hundreds that produced vehicles, from the 1890s through today, only a few remain. The attrition rate has fluctuated over the decades, the most dramatic occurring, not surprisingly, during the Great Depression. But even before Black Tuesday in October 1929, several firms had already disappeared: in one five-year period, from 1921 to 1926, a full 50 percent experienced mortality, leaving just forty-four companies in operation.

Like the ghost towns of the western frontier, the products of ambitious auto visionaries allude to what made America the world's industrial and commercial giant. These products—America's milestone automobiles—contribute a shared legacy, a collective vision of what the American automobile represented, from the vintage era of horseless carriages through the classic era of pre-Depression styling to the powerful, sleek, and ultra-efficient machines of today.

To quote the slogan of Henry Leland's Cadillac Automobile Company—manufacturer of the 1903 Cadillac Model A—American automotive inventiveness soon became the "Standard of the World."

FROM CARRIAGES AND CYCLING TO MOTORING

The sharing of ideas is not unique to the American automotive experience. What is unique is the sheer magnitude of the sharing. One innovation begat another in a surging sea of capitalism. Agents at the U.S. Patent Office were a busy bunch at the turn of the twentieth century, and applications for automotive grants largely contributed to the workload.

It is possible to explore the early American auto scene by following the likes of Charles B. King, a name often overshadowed by more famous pioneers, such as Henry Ford, Jonathan Maxwell, and Ransom Olds among others. It was actually King, the first to pilot a horseless carriage on Detroit streets, who mentored the likes of young Henry Ford. Ford, in fact, followed King that day in 1896 on his bicycle. It would be another three months before Ford test-drove his own powered carriage. "I am convinced that, in time, the horseless carriage will supersede the horse," King told the *Detroit Journal*. The early automobilist would go on to introduce the first vehicle with left-hand steering and center controls.

This volume is best served by beginning in 1896, three years after the appearance of the American Duryea brothers' first vehicle. It was during this period, just before and just after the turn of the century, that an explosion of assiduous creativity occurred in the American heartland, points east notwithstanding. By this time, of course, the gasoline-vehicle idea was making headway, following the pioneering work of Daimler and Benz in Germany in the mid-1880s. By the early 1890s, a number of American inventors had ventured into the field, among them John William Lambert in Pennsylvania, the Duryea brothers in Massachusetts, and Elwood P. Haynes with the Apperson brothers in Indiana.

Others, meanwhile, sought their fortunes in either steam or electric vehicles. "Steamers" proliferated during the antique-auto era, outselling gasoline reciprocating-engine cars by large numbers. Stanley, White, and Locomobile led the charge, but it was Stanley that hit steam's high watermark in 1910, after other steam-powered competitors had fallen by the wayside. Some, such as Doble, would later incorporate steam-powered models into their lineups, and do very well. Regarding manufacturers of electric automobiles, such as Detroit Electric, Riker, Columbia, Waverly, and Baker, battery technology was simply not advanced enough to further the idea. By 1920, early electric cars had gone the way of the buffalo.

In 1901, Olds Motor Works planned to market a lineup of gasoline and electric vehicles, all produced at its three-story factory in Detroit.

No matter the power source, turn-of-the-century vehicles were very much an amalgamation of bicycles and horse-drawn carriages. The "horseless carriage" was indeed just that, oftentimes constructed with bicycle frames and wheels. It is no surprise that the captains of the early American automobile industry were tenured managers, engineers, and innovators at carriage and bicycle manufacturers. Examples from both industries include Colonel Albert A. Pope, king of the American bicycle industry, and the Studebaker brothers, who ran the world's largest horse-drawn carriage company before venturing into the promising world of the automobile. The Winton Bicycle Company, founded by Alexander Winton and based in Cleveland, Ohio, is another clear example of this evolutionary link. Winton had turned his attention to autos as early as 1894, his first motor carriages rolling out two years later from the Brush Electric Company's factory.

Peerless, one of the "three Ps" (with Packard and Pierce-Arrow), was another bicycle manufacturer. Thousands of Peerless bicycles were made in the 1890s, but by 1900 the bike craze was over. Peerless was making parts for Winton and White, and transmissions for the French company De Dion Bouton, and considered getting into the car market itself. De Dion Bouton's engine was a good, reliable unit, used in Europe, and Peerless obtained a license to make machines under the firm's patents. Peerless used the De Dion Bouton frame, engine, and gear, but with its own style of aluminum body and bicycle wheels. The 3½-hp runabout was steered with a tiller, its rear-mounted engine allowing room for a box attached to the front of the dashboard that could be folded out for an extra seat for two passengers. The car weighed 700 lb (318 kg) and had a top speed of 25 mph (40 km/h). On November 1, 1900, the first

Peerless Motorette was displayed at America's first auto show in New York, advertised as the following year's model and priced at $1300. By 1903, although cars were being produced, the factory was still being used for bicycle production.

Ransom Eli Olds was also involved at this time. His first effort, also completed in 1896, was one of the pioneer gasoline automobiles in Michigan—indeed, in the entire country. In 1897, the Olds Motor Vehicle Company was established with a capital of $50,000, and the directors asked Olds to build one car "in as nearly perfect a manner as possible and complete it at the earliest possible moment." This was duly done. After producing a few cars, Olds moved from Lansing to Detroit in 1900. In 1901, a fire destroyed all the company's cars except for a little one-cylinder Curved Dash model. Olds resumed building Curved Dashes that same year, adopting the Oldsmobile name. The company's factory, built in 1899–1900, was reputedly the first in America specially designed for automobile production, and the largest of its day. From here, Olds put his famous Curved Dash model on the market, using the world's original assembly line.

Olds was a brilliant and bold man who, possibly more than anyone else, was responsible for starting Detroit on its course to becoming "Motor City." But he was still, in manufacturing terms at least, essentially a "farm mechanic." One of the major tasks in the factory was the hand-filing of transmission gears to make them mesh. This was costly, laborious, and time-consuming; in addition, the transmission as installed in the car was intolerably noisy. Olds went to a master of precision-made gears, Henry Leland—the eventual founder of Cadillac—for help, and Leland supplied him with a quiet-running transmission in which all the gears were precision-ground and interchangeable car to car without any hand-fitting.

Subsequently, in June 1901, Leland's firm was given a contract to make 2000 engines for Olds. Once again the man of precision demonstrated the superiority of his methods. The only other suppliers of engines to Olds were the Dodge brothers, John and Horace. Tests showed that the Dodge-built engine produced about 3 hp, whereas the Leland-built version developed 3¾ hp.

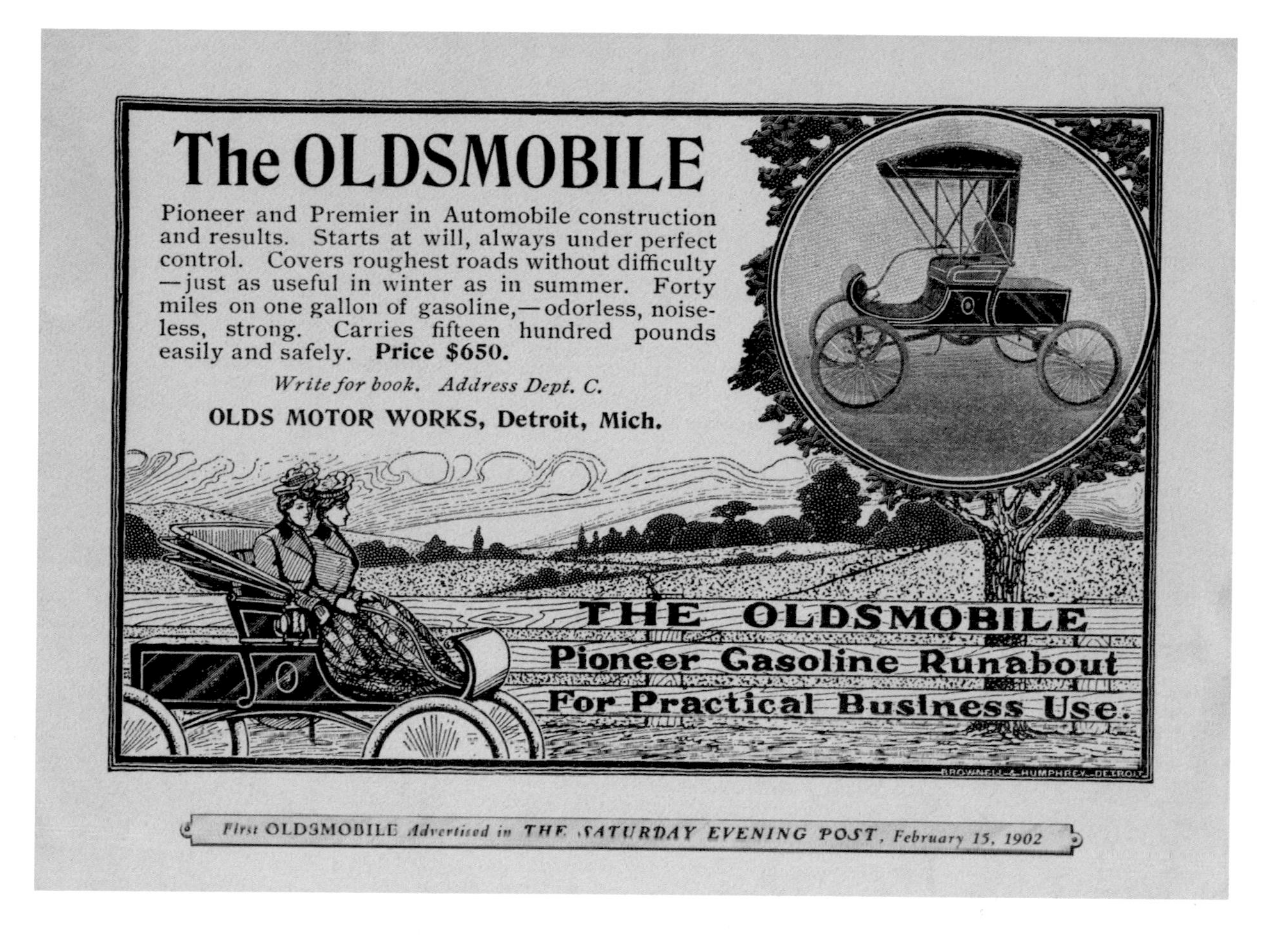

Ransom E. Olds built his first steam car in 1887, and his first gasoline model nine years later. The 1901 "Curved Dash" was the first vehicle to bear the Oldsmobile name. By 1904, more than 12,500 Curved Dashes had been sold.

Later in 1901, during the first automobile show to be held in Detroit, the Olds exhibit featured one of Leland's engines and a Dodge engine running side by side, dials indicating identical speeds for each. As Wilfred Leland, Henry's son, recalled: "Father and I stood looking at the dials, and a stranger said, 'Look behind that dial.' We saw a cheat brake load holding our own engine down to the same speed as the unbraked Dodge-built engine. It amused us. The man left. Some years later, when Henry Ford called to ask [father's] advice on grinding pistons, we recognized him as the man we had met at the stand."

With the brake load removed, the Leland-built engine ran at higher speeds than its Dodge counterpart. It had lower friction losses because of closer machining, and the result was enough for Henry Leland to realize his import to the nascent automobile industry. Cadillac was founded in 1902.

Also in Detroit, after moving from Warren, Ohio, brothers J. W. and William Doud Packard produced their eponymous marque's models of quality. By 1904, Packard had adopted the internal-expanding cone clutch—so too had Apperson—as seen in the impressive Packard Model L of that year. Attached to the engine's flywheel, the clutch utilized a propeller shaft to transfer drive to the three-speed transmission; this in turn combined with the final drive and differential gears in an extension of the rear-axle casing. This design was the forebear of the modern transaxle.

Detroit had not cornered the market in automotive inventiveness, however, especially prior to the reign of the "Big Three"—General Motors, Ford, and Chrysler—starting in the late 1920s. New England was also a hotbed of creativity, thanks in large part to the influence of Albert A. Pope and his Pope Manufacturing Company. A Bostonian, Pope held interests in several companies, not the least of which was Columbia. Based in Hartford, Connecticut, Columbia was a manufacturer of best-selling, early American four-cylinder autos available with both electric motors and gasoline engines. The Locomobile Company of America—first, at the turn of the century, a producer of steam cars, and later of gasoline-powered machines—operated out of Newton, Massachusetts. By 1902, when gasoline models were added to its lineup, Locomobile had completed about 5000 steam cars over a four-year period. Locomobile became well known for its stout, long-lasting products, one of which was the Type D prototype of 1904, still running after approximately 170,000 miles (273,588 km) in 1912 when the *Horseless Age* noted, "This car inaugurated many of the features which gained the Locomobile its reputation for strength and safety."

Such a reputation must have resonated with Henry Ford, father of mass production, who sought to deliver Everyman's car in his Model T (see page 47). In 1903, when Ford Motor Company opened for business in a barn-like building in Detroit, the automobile was an expensive plaything of a few wealthy motorists. Henry Ford envisioned and developed a vehicle that the common man could purchase, repair himself, and utilize for rural-life chores. Ford reportedly was once asked by one of his engineers how much space should be left between the Model T's front and back seats. Henry replied: "There should be enough room for a farmer's milk cans."

Pioneering the "moving assembly line" in 1913 was a masterstroke for Ford, one born, in fact, of necessity. Workers could now build a car in ninety-three minutes, compared to the days or even weeks required by other car manufacturers. In 1914, Ford promised to return between $40 and $60 to each customer who bought one of his cars during the following twelve months if sales during that period reached 300,000 units. The final count reached 308,213, and $15.5 million in $50 checks were duly returned. Proof of the assembly line's impact may be seen in the explosion in production: 6.7 million passenger cars left factories in 1919, compared to 23.1 million just ten years later.

It was also in 1914 that Ford turned the industry on its head when he announced his "$5-a-day workday." The prevailing wage at the time was $2.34 for a nine-hour day, and when the $5-a-day minimum wage (for an eight-hour day) was established, bankers and other businessmen thought he had lost his mind. Ford's contention that higher wages meant a better product—and thus better business—for the employer was proved right with time. The Tin Lizzie, as the Model T was known, was finally outmoded in 1927, by which time roads had improved to the point where the heavy-duty, high-riding T required a successor. The debut of its replacement, the second Model A, was described by historian Frederick Lewis Allen in his book *Only Yesterday: An Informal History of the 1920s* (1931) as "one of the great events of the year 1927, rivaling . . . the Lindbergh flight, the Mississippi flood and the second Dempsey–Tunney fight in its capacity to arouse public excitement." Indeed, when the car was shown at New York's Madison Square Garden, the size of the crowds broke all records for an indoor exhibition to date.

The American automotive industry made it through the early years of the First World War relatively unscathed. Uncertainties of a world economy in chaos were weathered, and despite rising costs of material, labor, and money, production rose dramatically and prices fell. The near-total sacrifices in private automobiling in Europe were mildly shared, if at all, in America. For the United States the conflict was short, and it ended before the nation's industrial war machine hit full throttle. What difficulties the war created in America were checked by the armistice. Ford closed Model T production only two months before the end of hostilities, and most factories maintained some

Opposite: In 1913, the Ford Motor Company introduced the first moving assembly line at its factory in Dearborn, Michigan. Workers no longer had to move from car to car to add their parts. Instead, the cars came to them.

Below: The first transcontinental auto race in the United States, from New York to Seattle, was held in 1909, and was won in a Ford Model T. The winners took 22 days and 55 minutes to complete the race, at an average speed of 7.8 mph (12.6 km/h).

William C. Durant was a charismatic businessman who, by 1907, had transformed the floundering concern that was Buick into the second largest and most influential auto company in the United States. He would use this leverage to form General Motors the following year.

automobile production throughout. However, when the country returned to business as usual, business was anything but.

SWINGING PENDULUM: FROM THE 1920S THROUGH THE 1930S

In September 1919, *The Automobile* noted that the following year's market demanded 2.5 million more cars than would normally be produced. Ford alone was projected to make 2 million. Willys-Overland planned on 200,000, two-and-a-half times its output of 1919. *The Automobile* reported that General Motors would invest $4 million in Olds Motor Works. Then reality tested the industry when the economy stumbled as American businesses tried to find their place in peacetime. Previously, the auto industry had endured most economic knocks without much pain. Now, however, commitments were being forfeited, from consumer to dealer to factory to supplier. The burden inevitably fell on creditors, who in turn tightened up. The financial community, which Henry Ford damned as "money sharks," called for liquidation of automobile debts; manufacturers, meanwhile, began to trump competitors by lowering prices, which were still well above the levels of 1917.

The automobile industry's time of trial was especially felt at General Motors Corporation, which had formed in 1908 with American businessman William C. Durant at the helm. Under Durant's intrepid—some say reckless—direction, GM was created through the acquisition of many small and medium-sized car companies, including Buick, Olds, and Cadillac. In the panic of 1920, Durant lost control of GM for a second and final time, on this occasion leaving to start his own company, Durant Motors. By then, GM consisted of rather wildly unrelated companies with little central control. In 1921, the firm's automotive segments garnered only 12 percent of the U.S. market, compared with Ford's 60 percent. But the roaring tide of automobile production came after the sharp economic recession of 1920–21, and it was GM that was most opportunistic. It would soon become the biggest of the Big Three.

In the mid-1920s, the American automobile industry entered the second stage of its history: that of competition rather than growth. Marketing outstripped production as the main issue to master, as it became paramount to induce consumers not to buy their first car, but to buy a new car.

It is at this point that the importance of styling arose in the American mindset. One of the main objectives of automotive development in the United States during the first quarter of the twentieth century was to ensure reliability and ease of use. Body-on-frame construction led to the rise of specialized automotive coachbuilders, who offered custom bodywork for wealthy motorists. Until the First World War, most cars were considered utilitarian objects. As tastes and wealth coincided, designers began to customize an automobile's body and produce artful, one-of-a-kind creations. Prefaced by this increased interest in style, the 1920s introduced the classic car, defined by the Classic Car Club of America as a model dated from 1925 through 1948.

The style-centric efforts headed by GM president Alfred Sloan first bore fruit in 1927, when GM finally usurped Ford Motor Company to claim the number-one spot with a U.S. market share of 43.49 percent, compared to Ford's 9.32 percent. (The tables had indeed been turned: just two years earlier, Ford was top with a 40.02 percent market share, GM with 19.97.) It is no coincidence that the turnaround occurred simultaneously with the creation of GM's internal Art and Color Section.

The first product of GM's styling department was the LaSalle of 1927 (see page 67), the first production car designed from the ground up by a professional designer, in this case Harley Earl, the department's head. The car's Hispano-Suiza look reflected Earl's knowledge of the most advanced European design. Until the LaSalle hit America, most domestic models were heavy and square, and rather somber. The LaSalle's smooth corners and graceful tablespoon fenders took the styling direction toward fleetness and grace, and a handsome radiator, combined with headlamps mounted on vertical stanchions and other details, set the car apart.

From there, the stylists and engineers at GM began to improve the overall appearance of the automobile and its components, paying attention for the first time to general visual impact. The main focus was on better proportioning, more pleasing lines, and a more harmonious blending of the car's individual components. Earl's objective was to seek new ways to lengthen and lower the American automobile, to "treat it not as an outgrowth of a wagon, but as an automobile with its own character and purpose."

Other domestic makers would soon follow the lead, and the American automobile came into its own. The direction of advancement was not confined to aesthetics, however. The convergence of forward-looking design and problem-solving engineering took place throughout the American automobile industry. After the First World War had ended, technology was king. Such earlier advances as the self-starter, developed by Cadillac in 1912, and the world's first V-12, Packard's Twin Six, were expounded upon and improved.

The great American entrepreneur E. L. Cord—the man behind the marque of the same name, as well as Auburn and, soon thereafter, Duesenberg—was the force behind the world's first front-wheel-drive automobile, the Cord L-29 of 1930. A few years later, in 1935, the Chevrolet came fitted with an all-steel roof known as the "turret top," a breakthrough made possible by new, big sheet-metal stamping machines; previously, car roofs had been made of crisscrossed wooden slats. The Oldsmobile of 1940 (see page 140) boasted the first fully automatic transmission, the Hydra-Matic; Cadillacs received it the following year. This advancement was the result of the solution to the problem of manually gear-shifting the then-primitive synchronization, which led to gear grinding. The automatic "slush box" (so nicknamed because of its fluid, rather than metal, connection to the engine) was a sign that American automakers were appealing to women drivers more than ever.

So admirable were the efforts of American ingenuity that French automobile creator and connoisseur Ettore Bugatti was the owner of a Packard straight eight, which influenced his Royale. Bugatti was a close follower of automotive advances in the United States, appreciating an industry that was well ahead of Bugatti's own engineering and standards of manufacture.

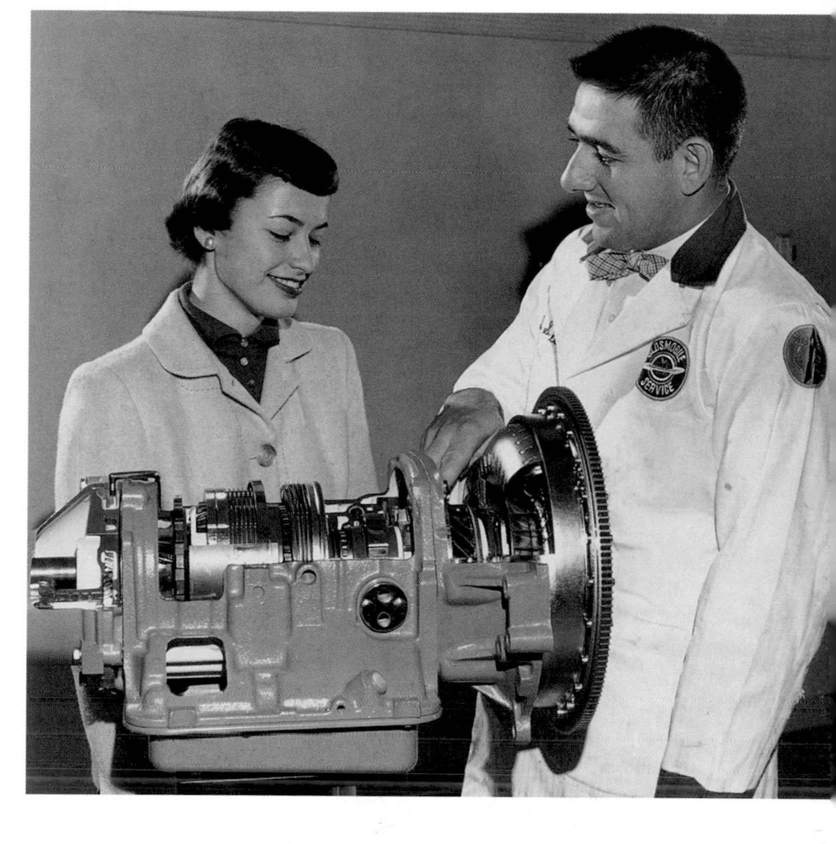
The evolution of the American automobile has been marked by advances in science and technology, including the introduction of the Hydra-Matic transmission on Oldsmobiles of 1940.

Opulence and luxury were the watchwords of the American automobile before the country entered the Second World War. Notable among the surge of prestigious makes was the Duesenberg Model J (see page 76), promoted as "the world's finest motor car." Duesenberg was part of E. L. Cord's business empire, and Cord instructed Fred Duesenberg to design "the biggest, fastest, and most powerful stock automobile the world has ever seen." Cord sought to capture the American buyer of such prestige brands as Minerva, Rolls-Royce, and Isotta Fraschini. He succeeded in late 1928 with the introduction of the Model J, which reportedly produced 265 hp and reached a test speed of 116 mph (187 km/h) at Indianapolis Motor Speedway. The chassis of the first Model Js sold for $8500, a fortune for the time.

Those automakers that survived the Great Depression drew on a combination of luck, deep financial reserves, and mold-breaking styling, as seen on the 1935 Nash.

In 1927, in addition to its famous Model A (replaced in 1932 by the Ford V-8), Ford Motor Company was manufacturing luxurious Lincolns. Ford had acquired Lincoln from Henry Leland in 1922, and the company's attention to high-quality workmanship continued in the cars' design and performance, traits very much of interest to Edsel Ford, son of Henry, who made the marque something of a pet project. Perhaps the greatest of all Lincolns was offered in 1933, the KB. The new V-12 engine generated tremendous acceleration and a top speed of 95 mph (153 km/h).

If Charles Darwin's theory of evolution were to be applied to the automotive industry, survival of the fittest would succinctly explain the demise of scores of carmakers in America after the Depression. No longer were roads traversed by new Auburns, air-cooled Franklins, or supercharged Duesenbergs. No longer were showrooms graced with current-year Pierce-Arrows, Marmons, or Stutzes. But others, such as Lincoln, Chrysler, and Nash, would ride on. Two of the Depression-era independent automakers that survived, Packard Motor Car Company and Studebaker Corporation, merged in 1954 to form Studebaker-Packard, which made automobiles until 1966.

To ensure GM's survival, Sloan conceived a business model he called "dynamic obsolescence." In short, it called for improving the design of a vehicle each year after its introduction until it became obsolete. The idea stuck. So much so, in fact, that all competitors were forced to follow suit when the car-buying populace demanded the latest cars—that is, next year's models. This became one of the hallmarks of postwar American car production. Futuristic design features decorated the new models from Detroit, Dearborn, and South Bend year in, year out.

The "classic" era of the 1930s featured the "teardrop" form, then considered the ideal aerodynamic shape. Streamlining, however, was coming into vogue. Examples of this new look included the Lincoln Zephyr from 1939 (see page 125) and the Chrysler/DeSoto Airflow models, which were also LaSalle's latter-day competitors.

"In the early 1930s, the variations in styling from year to year consisted of different treatment applied to the basic classic motif," explains designer Tom Tjaarda. (Tjaarda's father, John Tjaarda, designed the Lincoln Zephyr of 1935.) "The car of the future proposed in 1934, if looked upon today, resembles a somewhat streamlined version of a classic car. A new concept was needed in automotive research and design. This was answered partly in the Chrysler Airflow, with its truly functional seating package and ultra-soft independent front suspension. However, it was the Lincoln Zephyr with its unit body and rear engine (on the prototype), followed by the Continental with its beautiful styling [see page 135], that did much to revamp the whole automotive picture."

Company	Production units in 1941
General Motors	1,818,481
Chrysler	878,356
Ford	687,084
Studebaker	119,325
Nash	80,408
Hudson	78,117
Packard	66,906
Willys-Overland	28,935
Crosley	2289
Graham-Paige	544
Hupp	103

Source: Jan Norbye *et al. The American Car Since 1775*. New York: Automobile Quarterly, 1971.

GM's Harley Earl took the concept of streamlining one step further on a Buick chassis, creating the Y-Job of 1938. This one-off model, full of striking innovations in terms of both design and engineering, is thought to be the first car built strictly for promotional purposes. Granted, examples of one-of-a-kind show creations could be found at custom displays in the early 1920s at New York's Commodore Hotel. A $10,000 Judkins body on an equally valuable Duesenberg chassis was far from out of the ordinary. But show models produced from production cars were rarities, if indeed they existed at all, until GM's styling section began to create them in the mid-1930s. The car shows of that period—at Grand Central Palace in New York, at the Stockyards in Chicago, at the Convention Hall in Detroit, and at many other centers—were colorful and superlative. Part of their success in those years grew from a new concept developed by GM and its styling section for the shows of 1936 and beyond: to exhibit more at the exposition than buyers could see in the showrooms. Thus, the "show car" was born. It was still a production car, but the show car featured special paint treatment, special upholstery touches, and special handlings that were not in the order book.

For GM, show-model production became a standard year-by-year task after 1938, modified in one significant degree in 1939, when a Cadillac Sixty town car intended for display was made, beginning to end, in the styling section—not simply rebuilt there from a production-line assembly. Show-car development was married with the idea of experimental models, which would become the main attraction at GM's extravagant postwar Motoramas beginning in 1953.

At this point, it is interesting to note the diminished number of independent carmakers in the U.S. market as it stood in 1941, the last full manufacturing year before the United States entered the Second World War (see table, left). A full 90 percent of production was now in the hands of the Big Three, with four out of the eight independents producing a hardly measurable market penetration. All contributed to a shared mission,

Appealing to every sector of the market, General Motors was uniquely positioned to weather the Depression's storm. Cadillac's mid-entry-point offering was its "companion" car, the 1927 LaSalle.

Many auto legends were born in the years leading up to the Second World War, including the Duesenberg Model J, first seen in 1928.

however, when the country joined the war. A list of the industry's war-time contributions is testament to a manufacturing sector made for conquest: from the 207,400 pairs of binoculars to the 12.5 billion rounds of ammunition; from the 245 million artillery shells to the 4 million-plus engines; and from the 2.6 million trucks to the 27,000 aircraft, the extent of the industry's service was unmatched by any other industrial sector.

POSTWAR PROSPERITY LEADS TO FINS AND FINESSE IN THE 1950S

The "sleeping giant" awoken by the attacking Japanese in 1941 was powered by American automobile companies, which rapidly transitioned to producing military materiel. This accelerated the understanding and postwar use of lightweight materials, stressed bodywork, fuel injection, specialized production methods, and advanced aerodynamics.

In the booming economic years that followed the end of hostilities, the American auto industry geared up for mass production. Like their French counterparts, many of America's luxury car manufacturers shut down. Those that survived used lessons they had learned from the war, including the use of new lightweight materials, aerodynamics, and safety features, to produce cars that were more practical for both the consumer and the manufacturer. America's prosperity in the 1950s brought about a period of exuberance that manifested itself in stylish concept cars, many of which became production models. There was also a shift toward smaller, faster sports cars.

Hudson, which had converted to peacetime production in August 1945 and built 4735 cars by the end of that year, shared in the angst of the severe material shortages that characterized the immediate postwar period. While marketing only prewar designs for 1946 and 1947, it readied itself for the introduction of its advanced step-down design coupled with a new streamlined body. Although popularity of the design waned in the early 1950s, it did provide a new look and a highly capable road machine, and Hudson sales eventually matched prewar levels. The step-down design merged body and frame into a single structure, with the floor pan recessed between the car's frame rails instead of sitting on top of the frame. The first-generation Hudson Hornets (see page 163) utilized the design, and the resulting lower center of gravity helped the car handle well and win races. Hudson won twenty-seven of the thirty-four NASCAR Grand National races in 1952, followed by twenty-two out of thirty-seven in 1953, and seventeen out of thirty-seven in 1954.

Nash took until 1949 to offer its first all-new postwar cars, the unique Airflyte line of six-cylinder models, which, while ungainly from a design perspective, were known for their durability. In the meantime, the firm was quickly able to re-outfit its production lines and rise to third in terms of output in 1945, after Chevrolet and Ford. Much of Nash's later success was due to the Rambler, a pioneer compact that company president George Mason envisioned as a necessary postwar development, and the only small car of the early 1950s to achieve any kind of lasting success.

Oldsmobile developed the compact, short-stroke, 135-hp Rocket V-8, which it introduced in the 88 model of 1949. The Rocket was America's first high-compression overhead-valve V-8, and it made Oldsmobile the manufacturer of one of the world's best-performing cars, dominant in NASCAR events. Opinions on the origin of the American muscle car vary, but the Olds Rocket 88 cannot be denied its role as the first to respond to growing public interest in speed and power, putting a powerful engine in a lightweight Oldsmobile body.

Despite America's postwar boom times, starting a car company was difficult. The failure of Henry J. Kaiser and Joseph W. Frazer to do as much exemplifies the rough-and-tumble business environment, made even more unwelcoming by the Big Three's market dominance.

Henry Ford's grandson, Henry Ford II, found himself in an unenviable position when he assumed the company's presidency

Below: "Small," "sporty," and "elegant" were the watchwords of the auto industry's efforts to appeal to a new set of car buyers in the 1950s. Ford's offering was the Thunderbird of 1955.

Opposite: Cadillac's advertisements, such as this appeal to prospective buyers of the late-1950s Eldorado Brougham, were as glamorous as the offerings themselves.

in 1945 after VJ Day. The Rouge plant in Dearborn, Michigan, had quickly and successfully reconverted to peacetime production, and a general strike shut down GM in late 1945, which left only Ford and Hudson making cars at the close of the year. But Ford Motor Company was in trouble. Since losing its market-leading position during the Depression, little had been invested in plant facilities, and the sales staff had disintegrated during the war. Henry rose to the challenge, rebuilding the company from the inside out and its sales force from the bottom up, with fresh blood in key executive positions. The new decentralized management system—an unspoken concept within Ford before the death of patriarch Henry Ford in 1947—resulted in a modernized Rouge plant and, eventually, a thriving racing program.

After offering prewar leftovers to a car-starved postwar market for a couple of years, GM kicked off its peacetime rebirth in 1948 with a freshened Cadillac adorned at the rear with the industry's first tail fins (see page 155). A ground-breaking overhead-valve V-8 was dropped into the Cadillac of 1949, and a smaller version of this milestone mill found its way into Oldsmobile's first refashioned postwar model that year.

Though they still relied on prewar powertrains, both Buick and Pontiac also featured new façades in 1949, as did Chevrolet.

As America entered the 1950s, Chevrolet was determined to retain its prewar position as the top seller of domestic models, hitting the market with its promising new bellwether, the Bel Air. The Bel Air was rolled out in 1950 as Chevrolet's trend-setting "pillarless" coupe, a fashionable "hardtop convertible." It was quickly followed in 1951 by low-priced rivals from Ford and Plymouth, the Victoria and Belvedere coupes respectively.

The hardtop-convertible idea, first envisioned by Chrysler designers in 1946, combined the light, airy feel of a ragtop with the convenience of a conventionally enclosed automobile. The trick involved removing the vertical B-pillar that normally tied a traditional coupe's roof to its rear-quarter bodywork behind the doors. Chrysler's pioneering plan involved simply attaching a club coupe's roof to a Town and Country convertible. Seven such prototypes were created in 1946, but regular production did not follow. Independent Kaiser toyed with the idea as well, by welding a steel top on to its somewhat odd four-door convertible in 1949. This effort was also unfruitful, however, and it was left to GM, Detroit's styling leader, to popularize the hardtop-convertible concept.

Cadillac's Coupe de Ville, Buick's Roadmaster Riviera (see page 157), and Oldsmobile's 98 Holiday all debuted with the hardtop-convertible look in early 1949. Sales of these high-profile two-door coupes were initially limited, but the numbers were sufficient to convince GM planners to promote the platform further. GM's hardtops entered the high-volume realm the following year, after Buick and Olds introduced lower-priced variations on their Riviera and Holiday themes, and both Pontiac and Chevrolet joined in with their new Catalina and Bel Air coupes respectively.

Performance expectations continued to challenge engineers as well. In 1955, Chevrolet introduced the GM small-block V-8, bringing to everyday cars the high-revving performance once reserved for exotic luxury machines.

Manufacturers across the board showcased their hardware in flashy limited-edition models. Chrysler led the way with its C-300 of 1955, a blend of Hemi power and luxury-car trappings. Producing 300 hp, it was advertised as "America's Most Powerful Car." With a 0–60 mph (97 km/h) time of 9.8 seconds and a top speed of 130 mph (209 km/h), the C-300 was one of the fastest asphalt eaters of its day.

The demand for light sports cars hit America at this time. In terms of such vehicles, the country was lagging behind others, particularly Britain. GM's answer was the Chevrolet Corvette (see page 170). The public's response to the Corvette at the New York Motorama of January 1953 was favorable, even if it was only in relation to the new roadster's styling; by European sports-car standards, it was underpowered. Six months after this debut, the first production Corvette left the factory gates. Fiberglass was chosen as the body material not only for its weight-saving properties but also because it gave the designers greater freedom and made tooling easier to create; the rapid production start would not have been possible otherwise. In July 1953, Chevrolet chief engineer Ed Cole received expert reinforcement in the form of Zora Arkus-Duntov, a young engineer and motorsports enthusiast. The Corvette has always stood apart in terms of styling, as well as, eventually, performance. The chrome teeth of its grille grinned like a predator, and the front was further defined by headlights under a gravel guard. Passengers in the two-seater were protected by a panoramic windshield. On the performance side, in 1955 Cole and Arkus-Duntov replaced the existing six-cylinder with GM's small-block V-8, which, initially, had a capacity of 262 cu. in. and produced 195 hp. The Corvette actually became a sports car in its own right in 1957, the first year it was offered with a four-speed transmission and fuel injection, making it the modern American sports car.

By the mid- to late 1950s, automotive production was finally able to meet the needs of a postwar, car-hungry public. The emphasis had shifted from acceptance of any available new car to a sudden interest in quality and technical innovation. At the time,

The 1960s witnessed radical changes not only to social relations but also to fashion. The car industry was not immune, as demonstrated by the 1963 Corvette Sting Ray (above) and the 1965 Ford Mustang (opposite).

the latter consisted of what became known as the "Space Race," which, combined with a national milieu of unfettered optimism, explains the design direction that defined the decade.

The idea for tail fins probably has its origins in GM design chief Harley Earl's fascination with the Lockheed P-38 Lightning fighter aircraft, which he had observed just before America's entrance into the Second World War. The plane's streamlining and individuality stamped an indelible impression on Earl and others, and facets of the P-38's twin tails and booms were incorporated into later postwar designs, most noticeably in the form of Cadillac's "fish tails" of 1948. The standoffish attitude of many Cadillac dealers, fearful of a design lemon, soon changed when sales responded superbly; buyers related the new treatment to prestige and distinctiveness.

Cadillac's fins attained their zenith on models of 1959 (the development of which began in late 1956; see page 205), reaching 22 in. (56 cm) in height. All other GM divisions got in on the act, notably Chevrolet with its Bel Air (see page 197), Buick with its Invicta, and Oldsmobile with its Super 88. The flashy Chevrolet Impala of 1958 was treated to no fewer than six taillights; the following year saw the Impala at the top of Chevrolet's lineup.

Chrysler sought to outdo GM in the quest for outlandish styling treatments. Virgil Exner led the Chrysler design team in its use of fins, which became a major design feature for 1956 and increased in size on an annual basis. The bolted-on appearance of Chrysler's fins gave way to a more integrated design on the New Yorker, Imperial, and 300-C. Plymouth fins reached their apex in 1960 on the Belvedere and Fury.

The battle for new fin heights raged among all parties. In 1955, Ford launched the Crown Victoria, which featured a blade-like fin atop the rear quarter panel; the next four years saw the company apply more obvious treatments to its Thunderbird (see page 185) and Fairlane. Others to sport fins included American Motors (the product of a Nash–Hudson merger) and Studebaker.

By the end of the 1950s, Bill Mitchell, Earl's successor, and others were on a quest to take design down a more "slender" path, with less reliance on fins. Beginning with its models of 1960, which had been in development since 1958, GM moved away from the excesses of the previous decade and toward more sharp-edged designs. Again, other automakers followed suit.

FLEXING THEIR MUSCLE: MUSCLE CARS IN MOTOR CITY

The American car industry in the 1960s was far removed from the "decade of love"; indeed, there was no love lost between Detroit, where GM and Chrysler staged battle, and Dearborn, where Ford developed its war plans. At Ford, Lee Iacocca—who would later become Chrysler's knight in shining armor—knew that the company needed a youthful, affordable car even before

he was made vice president at the age of thirty-six in November 1960. "We approached the decade of the 1960s with a rather stodgy, non-youth image," he said. That image would change with the arrival of a sporty little number called the Mustang.

Iacocca set the tone for the Mustang: it had to be small, light, and inexpensive—no more than 180 in. (457 cm) in length, 2500 lb (1134 kg) in weight, and $2500 in price—yet capable of carrying four people. Its styling would employ the sporty long-hood, short-deck, low-profile look that had made the two-seat Thunderbird a modern classic. It would be available with either a six-cylinder or a V-8 engine, and would be highly adaptable in order to suit a wide variety of tastes. In short, it would have the flair and performance of a Thunderbird at the price of a compact Falcon.

"With a two-door sporty coupe, you've got to shock them a bit at first," Iacocca asserted. "Even when it seems you have a certain winner, people are going to worry just because it's different." What worries did exist were overcome by reality: the

GM president Ed Cole in 1967 celebrating another landmark in the company's remarkable history. The 1966 Olds Toronado (also opposite) was a *tour de force* in styling and engineering design.

1964½ Mustang was an unprecedented hit, selling a record 418,812 units in its first twelve months on the market, with sales reaching 1 million in less than two years.

The 1960s saw American car companies react to the popularity of a number of foreign makes. The Corvair of 1960, for example, was Chevrolet's answer to Volkswagen's success. But drivers of the rear-engined Corvair were losing control of the Euro-style car and crashing; drivers could overturn the vehicle just by going around a corner if tires were slightly under-inflated. The Corvair became the poster child for how not to compete with imports—that is, rushing development at the cost of safety—and was the subject of a chapter in Ralph Nader's landmark consumer book *Unsafe at Any Speed* (1965).

Consumers saw a different image in print and television advertisements, of course, as ad agencies chosen to run extensive campaigns did more than sell the American Dream to car buyers. The concept of entitlement filled the colorful ad space, successfully emotionalizing reasons to buy through clever copywriting and artful rendering. Those agencies with the most talented artists won the biggest accounts, and one team in particular stood out in the 1960s: the prolific duo of Art Fitzpatrick and Van Kaufman, the former illustrating the model of the year, and the latter producing an intriguing background scene. The ads went beyond empathizing with the reader by placing them in exotic locales and in the car. "This could be you!" the ads screamed; what could be more American than that notion?

Fitzpatrick, whose artwork has been commissioned by thirteen different automakers, spent twenty-one years at GM, beginning with the ad campaign for the Buick of 1954 (see page 181). He and Kaufman are well known for their work on Pontiac campaigns of the 1960s. A series of six ads devised for Pontiac's new Grand Prix in 1962 featured loose paintings with plenty of line drawing and simple backgrounds. The announcement ad, showing the Grand Prix and race cars in a Monte Carlo setting, combined with spirited text, attained the highest Nielsen rating Pontiac had ever achieved. More importantly, these six magazine ads alone—without the help of television or newspaper advertising—established the Grand Prix as Pontiac's new personal luxury car, ousting Ford's Thunderbird from the top of its class.

Perhaps more poignant in terms of impact was Pontiac's other popular model, the GTO (see page 216), originally an option package featuring Pontiac's 389-cu.-in. V-8 engine, floor-shifted transmission with Hurst shift linkage, and special trim. In 1966, the GTO became a model in its own right, spearheaded by Pontiac division president John DeLorean. The new model

proved more popular than expected, and inspired GM and its competitors to produce numerous imitators.

For 1964–65, Mopar (Dodge, Plymouth, and Chrysler) unveiled the 426-cu.-in. Hemi engine, further escalating the race to the top of the power pyramid. A volley of competitors responded, including Chevrolet with its Chevelle Malibu SS, Ford with its Thunderbolt 427, and Buick with its Skylark Gran Sport.

A latecomer to the horsepower wars, AMC managed to generate legitimate drag-strip contenders beginning in 1965, when the company rolled out its Rambler Marlin fastback. The Marlin did not fare well in terms of sales and initial performance, but AMC gained some muscle-car credibility in 1967, when both the Marlin and the "more pedestrian" Rebel were offered with the new 280 hp, 343-cu.-in. Typhoon V-8. The following year, AMC offered two big-hitters: the Javelin, and a smaller version, the AMX.

In 1967, Chevrolet launched the Camaro and its Pontiac sibling, the Firebird. The Camaro was brought out to compete with the Mustang, as Chevrolet product managers made clear at the press debut when they were asked, "What is a Camaro?" Their answer: "A small, vicious animal that eats Mustangs." Not to be outdone, Ford set out to refine the car first introduced as a mid-year 1964 model, releasing a more streamlined Mustang in 1969 with an improved, 302-cu.-in. engine.

Other developments were occurring outside the muscle-car maelstrom, of course. Oldsmobile stunned the industry in 1966 with its front-wheel drive Toronado (see page 229), conceived as a full-size personal luxury car to compete directly with the Ford Thunderbird and Buick Riviera. With the Toronado, Oldsmobile succeeded in reviving front-wheel drive in the United States not used in America since the Cords of the 1930s—and laid the groundwork for GM's mass changeover to the set-up in the 1980s. (Following the GM trend, Oldsmobile gradually phased out rear-wheel drive, last seen on its Cutlass of 1990.) The Toronado was one of the most important models of the 1960s, and its "wheel hop" styling innovation is still used today.

AMERICAN CAR IDENTITY CRISIS

The horsepower wars continued into the early 1970s. Plymouth made its mark with the Rapid Transit System, a marketing campaign that introduced to the car-buying public the ever-beefy Barracuda, Road Runner, GTX, and Superbird (see page 248) models. The Dodge Challenger debuted in 1970 as a spinoff from the Plymouth Barracuda of the 1960s, and other models were continued, such as Ford's Mustang Mach I and Boss 302 Mustang, the Oldsmobile 442, and the Chevrolet Nova SS. But the end of muscle was nigh.

Many factors contributed to the end of the muscle-car era and the beginning of an American car identity crisis in the turbulent 1970s. The automotive safety lobby decried offering powerful cars for public sale, particularly when targeted at young buyers. It is true that safety concerns were commonly displaced by a desire to achieve maximum power: the muscle cars' muscle overwhelmed brake efficiency, handling, and tire adhesion. The automobile insurance industry responded by levying surcharges on high-powered street models, an added cost that put muscle cars beyond the reach of their target market.

Simultaneously, efforts to combat air pollution forced Detroit to focus on emissions control. The majority of muscle cars were equipped with high-compression powerplants, some with a ratio as great as 11:1, and 100-octane fuel was desirable and easily accessible—that is, until the passage of the Clean Air Act of 1970, when octane ratings were lowered to 91, thanks in part to the removal of tetraethyl lead as a valve lubricant. Unleaded gasoline was phased in from that point on.

Performance became paramount during the mid- to late 1960s. Pontiac's GTO was inspired by the Tempest Le Mans.

In 1973, OPEC, in alliance with Egypt and Syria, triggered an embargo on oil exports to the United States in response to the country's support for Israel during the Yom Kippur War. Oil prices skyrocketed, and gas rationing ensued. President Nixon asked gas-station owners not to operate on Saturday nights and Sundays. Lines formed at the pump. The situation flared up again with the energy crisis of 1979 during the Carter administration.

The U.S. government mandated that a 27.5-mpg (8.6 l/100-km) fleet average be achieved by 1985. By necessity, car stylists virtually wiped their drawing boards clean to create smaller, more fuel-efficient models. Much bemoaning incurred in Motor City, as well as in the automotive press. For a public content with heavy and comfortable transportation, being force-fed Ford Pintos and AMC Gremlins and Pacers came as something of a shock. Yet, not all acknowledged the sad state of styling. "I think the revolutionary changes in automotive design that the regulations have initiated are creating a great new market for our products," said Philip Caldwell, president of Ford Motor Company.

Whatever one's attitude might have been, Detroit stopped making the automobiles consumers had been used to seeing and driving. Horsepower ratings plummeted as engine compression ratios were reduced, high-performance engines phased out, and all but a handful of models discontinued or transformed into "sport" packages, such as Plymouth's Road Runner, an upscale decor package for Volare coupes. One of the last muscle cars to fall was Pontiac's Trans Am SD455 of 1973–74.

In addition to the forced downsizing of the American automobile, styling and engineering gaffes plagued the industry.

Muscle remains important to American auto enthusiasts, hence the recent roll-outs of several power packages, including Chevrolet's 2010 Camaro.

Just one example is the Oldsmobile diesel V-8 of 1978. During the oil-embargo era, in its effort to achieve better fuel economy, Oldsmobile converted a gasoline engine to diesel, but the switch was made without beefing up some key components, resulting in mass failures.

The Japanese-car invasion exacerbated American auto-industry ailments. Consumer demand seen in California for at least a decade—for smaller, fuel-efficient imports—had become a nationwide phenomenon. Up to that point, progressive dollar devaluations and overseas inflation had placed the foreign make out of the "economy class." Presented with domestic double-digit inflation, a nagging recession, and ever-rising purchase and ownership costs associated with homegrown vehicles, consumers opted for the only alternative: the growing supply of imports. By 1977, Toyota, Datsun, Volkswagen, and Honda held 18 percent of the expanding small-car market.

Another changing reality was the shift in brand loyalty. What was once a given—that once a Buick lover, always a Buick lover—was now a variable. American automakers were faced with the fact that new-car buyers, especially head-of-household women, tended to try unfamiliar nameplates with foreign names, switching from one to another, perhaps for the sake of adventure. Consumer demands seemed to fly in the face of Alfred Sloan's theories of planned obsolescence and trading up. As the 1980s dawned, imports snared more than one out of every four new-car sales nationally. Yet, as Sloan remarks in his memoirs, "the hardest learned lessons are the best learned ones."

REEMERGENCE AND RESURGENCE

What had made the American automobile great? This question has occupied the minds of U.S. automakers from the mid-1980s to the present day. The quest for producing what the American driver wants is responsible for the shift toward better design and improved engineering, some of it borrowed from the past, and some of it based on the ideas of foreign competitors.

In 1997, reported *Automotive News*, the U.S. luxury-car market was "flying high." Luxury-brand sales were up

In the uncertain market of recent years, American carmakers have had to differentiate marques and models as never before. Chevrolet ably rose to the challenge with its 2008 Corvette.

14 percent. The factors fueling the growth were a strong economy, baby boomers reaching their affluent years, newer products, and steady or lower sticker prices. Among the manufacturers in the luxury class, Cadillac held the most excitement. At a briefing in Carmel, California, in 1998, Cadillac confirmed earlier rumors that it was designing a new two-seat concept car. The resulting vehicle was revealed at the North American International Auto Show (NAIAS) in Detroit in January 1999, and came to production reality during 2003. It was reported to have cost Cadillac approximately $4 million to produce. As presented at the NAIAS in January 2000, the car was called the Evoq, and its purpose was explained by Wayne Cherry, GM's vice-president of design: to initiate Cadillac's new "Art and Science" product philosophy.

Cherry explained that future Cadillac styling would be "dramatically different." The existing "sculpted organic forms," as seen on the Catera and the De Ville, were to be superseded by computer-generated geometric shapes with sharp lines, flat planes, and crisp intersections. The new direction has influenced the look of products coming from GM's top division ever since. Although the corporation has struggled to survive—GM discontinued America's oldest automobile maker, Oldsmobile, in 2004, and Pontiac and Saturn in 2009—Cadillac, Chevrolet, Buick, and GMC remain.

One bright spot on the American auto horizon is a return of sorts to yesterday's legends. Pop culture has celebrated the Camaro, Mustang, and Charger, with the recent redesigns of these cars fueling their popularity even more. In 2005, an accolade for the much-loved designs of the 1960s was seen in an all-new, heritage-designed Mustang on a superior platform derived from Ford's Jaguar S-Type/Lincoln LS architecture. Enthusiasts did double-takes, recognizing the styling cues from

Mustangs of the late 1960s and early 1970s, most notably the jeweled, round headlamps in a trapezoidal housing. The Chevy Camaro, the most recent heritage-based design to resurface—in the form of a new 2010 model, production having ceased in 2002—is one of the biggest-selling cars in the United States.

Reflecting the extent to which the American car industry has changed since the 1980s, when Lee Iacocca attempted to build ramparts to thwart a "Japanese invasion," 83.8 percent of all Honda and Acura automobiles sold in the United States in 2009 were produced in Honda's North American plants, an all-time high. The company has maintained a local production rate of 75 percent or higher since 1996, with no fewer than four manufacturing centers in the United States. It can no longer be said that Hondas are a purely foreign product. Mirroring the presence of American automakers' production facilities and sales in other countries, the great melting pot of American automobile manufacturing—including Honda, Toyota, Hyundai, Volkswagen, and others—completes the cycle of globalization.

Another cycle is currently being completed, that of the return to small cars. This earnest movement is reminiscent of the death of the muscle car and harried introduction of the compact car during the fuel crisis of the 1970s. One difference is the available and developing technology that innovators are adopting to raise fuel efficiency and lower fossil-fuel dependency. Electric and fuel hybrids are now commonplace on American roads.

Clearly, business as usual in Detroit was usurped when GM and Chrysler went bankrupt in 2009. Their dire financial straits were the result of not heeding observers' calls, including those from President George W. Bush in 2007. The Big Three were "going to have to learn to compete," said Bush. The learning curve abruptly ended soon after that statement. Now, as a major stakeholder in both companies, the U.S. government, under President Barack Obama, is ushering in a new era of mandated, eco-friendly changes. Amid the inevitable grumbling within the industry, a general sense of acceptance—if not acquiescence—is pervasive.

The NAIAS of 2010 highlighted the new small cars soon to be seen in America. Many carmakers have hedged their bets with these little four-cylinder compacts and subcompacts, some of which can achieve 40 mpg (5.9 l/100 km) on the highway. These new models prove that "cheap and cheesy" compacts are a thing of the past. The 2012 Ford Focus is a case in point, with power options including a 112-cu.-in. gasoline engine and Ford's Ecoboost four-cylinder unit. The new Focus will provide luxurious amenities and high-tech features in a sedan or four-door hatchback body, a stylish exercise in the new era of American motoring.

Although it is unlikely that annual U.S. car sales will return to the 17-million level of 1999–2007 for quite some time, American manufacturers are cautiously optimistic about the future. Ford and GM are producing more "global" cars, such as the Focus, which are constructed and sold around the world using common parts, so that a car built in China is almost identical to one built in the United States. The hope is that stylish innovation will continue to evolve—and produce American automotive legends for years to come.

This hope is shared by car enthusiast and comedian Jay Leno. "There are a lot of modern collectibles now," Leno explained. "I think first-generation hybrids like the Insight and early Priuses will have a nostalgia factor for young people, a nostalgia that we don't have. And they'll collect them."

There will always be room for another legend.

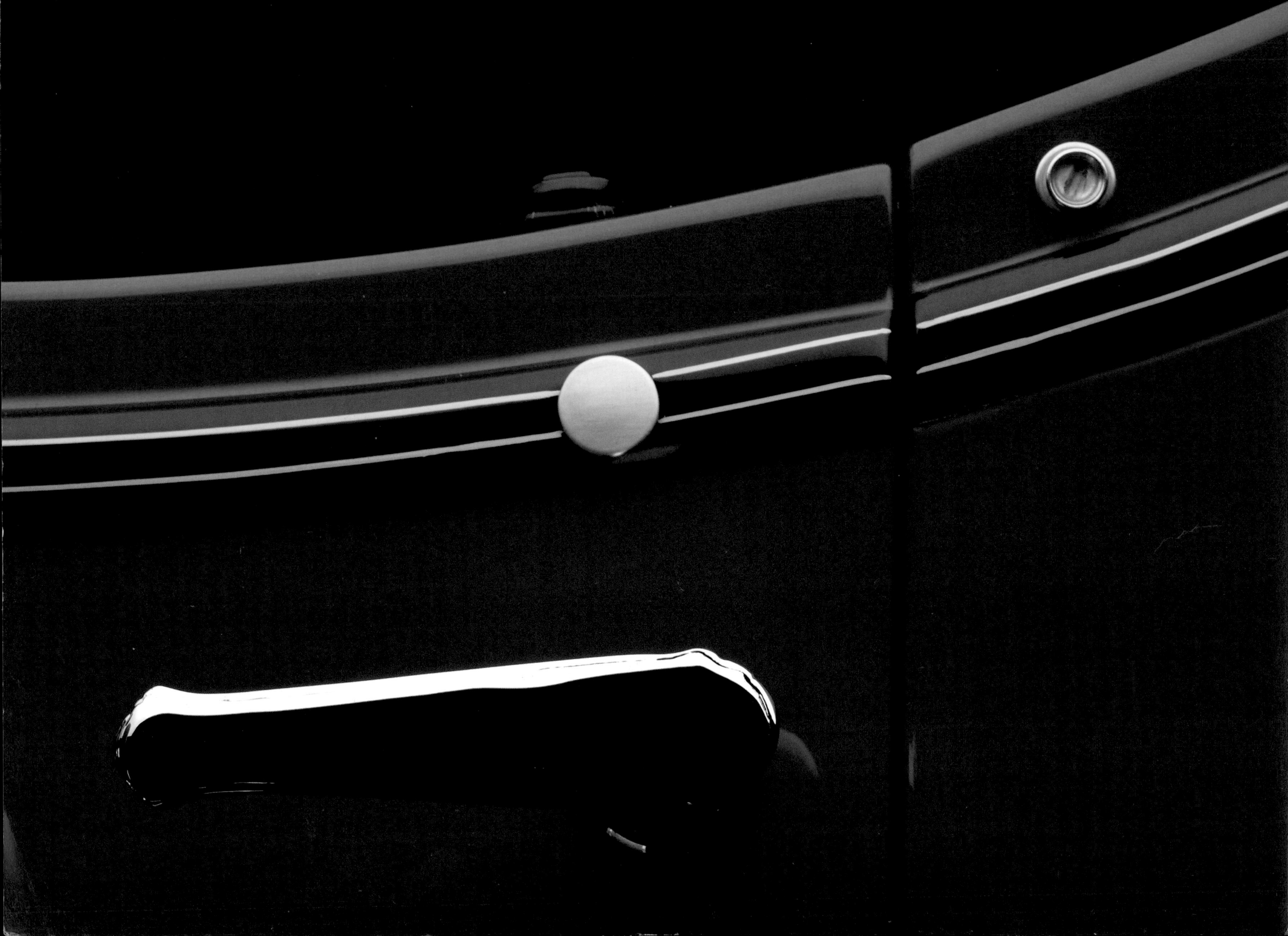

AMERICAN AUTO LEGENDS

1903 CADILLAC MODEL A REAR-ENTRY TONNEAU

Cadillac's first production-year model, originally coined simply "Cadillac" and later the Model A, was a single-cylinder automobile powered by the "Little Hercules" engine built by Leland & Faulconer, the engineering firm of Cadillac's founder, Henry Leland. Debuting in January 1903 at the New York Automobile Show, the Model A signaled to the burgeoning American auto industry that interchangeable parts were the way of the future. Leland was a master of precision, a perfectionist who demanded stringent machining and refinement of gears and castings. This quest for perfection bore early fruit when the first Cadillac to be exported to England—a Model A—was entered in the RAC's One Thousand Mile Reliability Trial of September 1903 and finished first in its price class on reliability scoring.

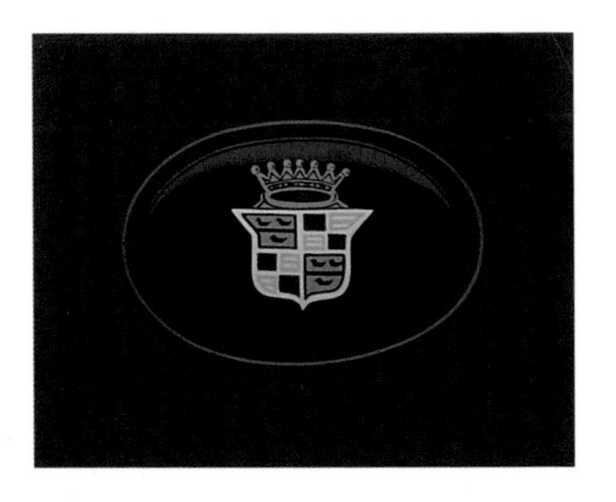

The single-cylinder, 5-hp Model A was the Cadillac Automobile Company's first vehicle, debuting in 1903.

The Model A was offered as a runabout for $750, or $800 with a leather top; the four-passenger tonneau cost $850. By early 1903, Cadillac had received orders for 2286 cars, a tremendous number. By mid-week of the New York auto show at which it was launched, Cadillac's sales manager, William E. Metzger, announced that the firm had "sold out," and no more orders were accepted.

These luxury vehicles of their day were distinguished by patent-leather mudguards and a steering wheel rather than a tiller. All Model As were painted red, weighed 1320 lb (599 kg), and sat atop a 76-in. (1930-mm) wheelbase. Top speed was a respectable 30–35 mph (48–56 km/h), with the engine rated at 5 hp. Final drive was by chain to the rear wheels, lubricated by beef tallow, as the factory recommended.

At the time of these first Cadillacs, Henry Leland was sixty years old, but age deterred neither the man nor his company's direction. With the Model A—the firm's entry into the American automotive fray—Cadillac had already set a new standard for refinement.

Nearly 2500 Model As were built in 1903, an exceptional achievement for first-year production.

The Model A was offered with the option of a detachable rear-entry tonneau body. Its maximum speed was 30–35 mph (48–56 km/h), while fuel consumption was rated at 25–30 mpg (8–9 l/100 km).

1903 STEVENS-DURYEA STANHOPE

Of the two Duryea brothers, Charles and J. (James) Frank, the latter was the conservative one, a trait that carried over to J. Frank Duryea's straightforward, no-nonsense automobiles. The brothers share the honor of producing America's first successful gasoline-powered automobile for commercial sale, which first ran on September 21, 1893, in Springfield, Massachusetts. But the brothers fought bitterly between themselves—a substantial disagreement being which of them could rightfully claim to have first developed their pioneering model—and they parted company in 1900. Frank ventured out on his own and formed the Hampden Automobile and Launch Company. Seeking more capital for his small manufacturing concern, he contracted with the J. Stevens Arms and Tool Company of Chicopee Falls, Massachusetts, to build, under his supervision, a Stevens-Duryea automobile.

Stevens-Duryea models were the result of a collaboration between Frank Duryea and the J. Stevens Arms and Tool Company. The first prototype appeared in 1901.

J. Frank Duryea had built three automobiles before making arrangements with Stevens, using the same Duryea opposed twin-cylinder engine he had developed in 1897. Now, moving his company into the larger Stevens facility, Duryea purchased the rights to his engine from his brother's Duryea Wagon Company for $2500; at the same time, Stevens began to manufacture several hundred engines. The production total for 1903 was 483 automobiles, and the models were known for their superb engineering. The Stanhope body style is characterized by its single, centrally mounted bench seat, its folding cloth top, and a "dashboard" at the front. All Stanhopes featured tiller steering, either in the center or at the side.

A total of 10,000 Stevens-Duryeas had been built by the end of the company's life in 1927. Duryea had withdrawn from active involvement in 1914 when he became ill, selling to Westinghouse Electric and Manufacturing Company the following year. Upon his retirement, he said he "saw little future in high-quality cars in the face of competition from mass-produced automobiles." Duryea died in 1967.

Left: The two-cylinder, 5-hp Stevens-Duryea runabout of 1903 was offered for $1300. The car was started from the seat using a hand crank.

Opposite: Lantern manufacturer R.E. Dietz Company introduced the first kerosene lamps for automobiles in 1896; by 1903, they were regarded as standard equipment.

DIETZ

1908 BUICK MODEL 10 RUNABOUT

In late 1907, after three years under William C. Durant's steerage, Buick was headed for the top tier of automakers. The frenzied acquisition of companies that would eventually form General Motors was already underway, and Durant's plan was to position Buick as the financial cornerstone of his empire. Utilizing his Durant-Dort Carriage Company's national dealer network, he quickly cobbled together the best system of automobile dealerships in America. Virtually overnight, Durant turned Buick into the country's most visible automotive concern simply by changing the names of his dealerships, and the Model 10 was the primary beneficiary.

The Model 10 was unveiled at the New York Automobile Show in November 1907, to the praise of *Motor World*, which called it the "sensation" of the show. Buick's charming, small, and affordable ($900) car was described by the company as "a gentleman's light four-cylinder roadster." Buyers knew the Model 10 as an automobile that was easy to operate, with a smooth-running engine and a big bang for the buck: included were acetylene headlamps, an oil lamp at the rear, and a bulb horn. The three-passenger car was highlighted with plenty of brass and handsome lines. The popularity of the Model 10 almost outpaced that of all other Buick models of 1908: 4002 Model 10s were sold that first year, making up nearly half of the firm's total annual sales.

Buick claimed that it was the world's number-one automobile builder in 1908, producing more than the next two competitors—Ford and Cadillac—combined. This is contested by some historians, however, who cite Ford's production total for its fiscal year rather than its calendar-year production. It is duly noted that Ford's Model T was not introduced until very late in 1908, thus tipping calendar-year production in Buick's favor. Whatever the case, it was the well-built and well-timed Model 10 that catapulted Buick to the top.

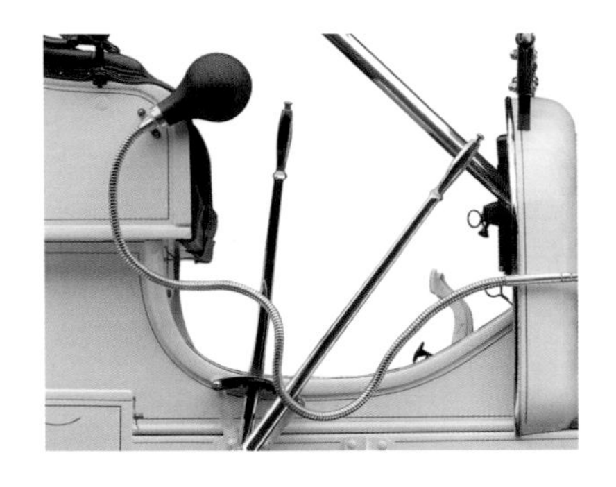

Finished in off-white Buick Gray and trimmed in brass, the Model 10 was Buick's most popular automobile in 1908.

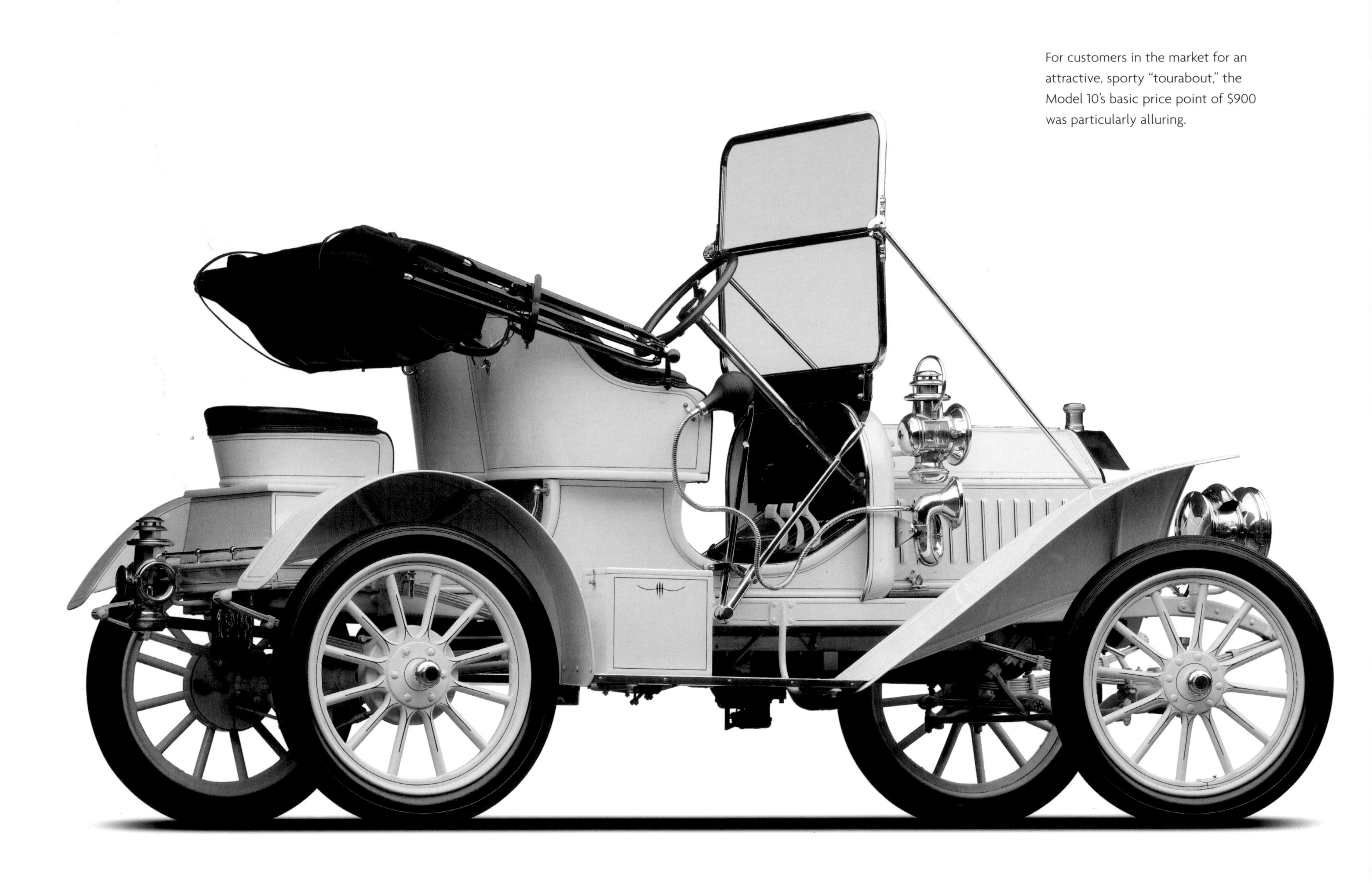

For customers in the market for an attractive, sporty "tourabout," the Model 10's basic price point of $900 was particularly alluring.

Buick's Model 10 Runabout came with a bulb horn, acetylene headlamps, and an oil-powered taillight as standard. More than 4000 of these three-passenger cars—powered by an in-line valve-in-head four-cylinder engine—were produced for 1908.

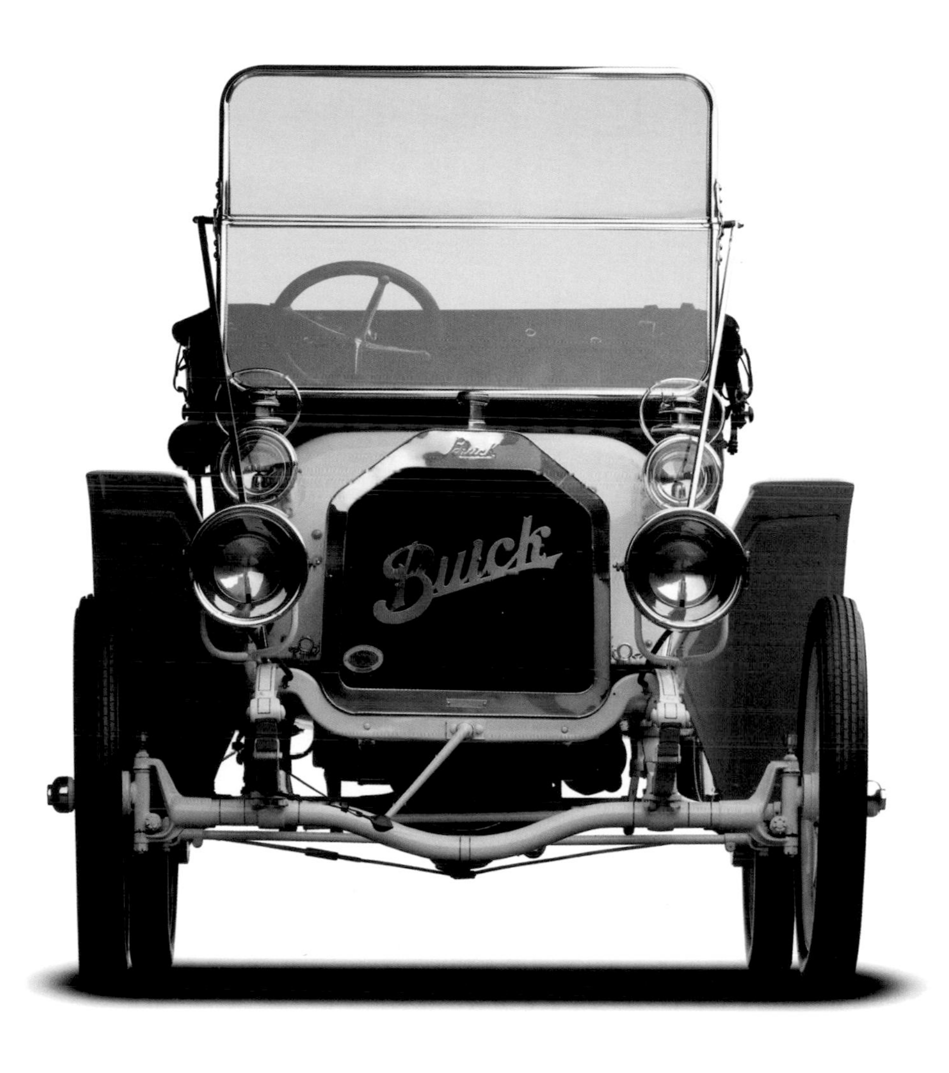

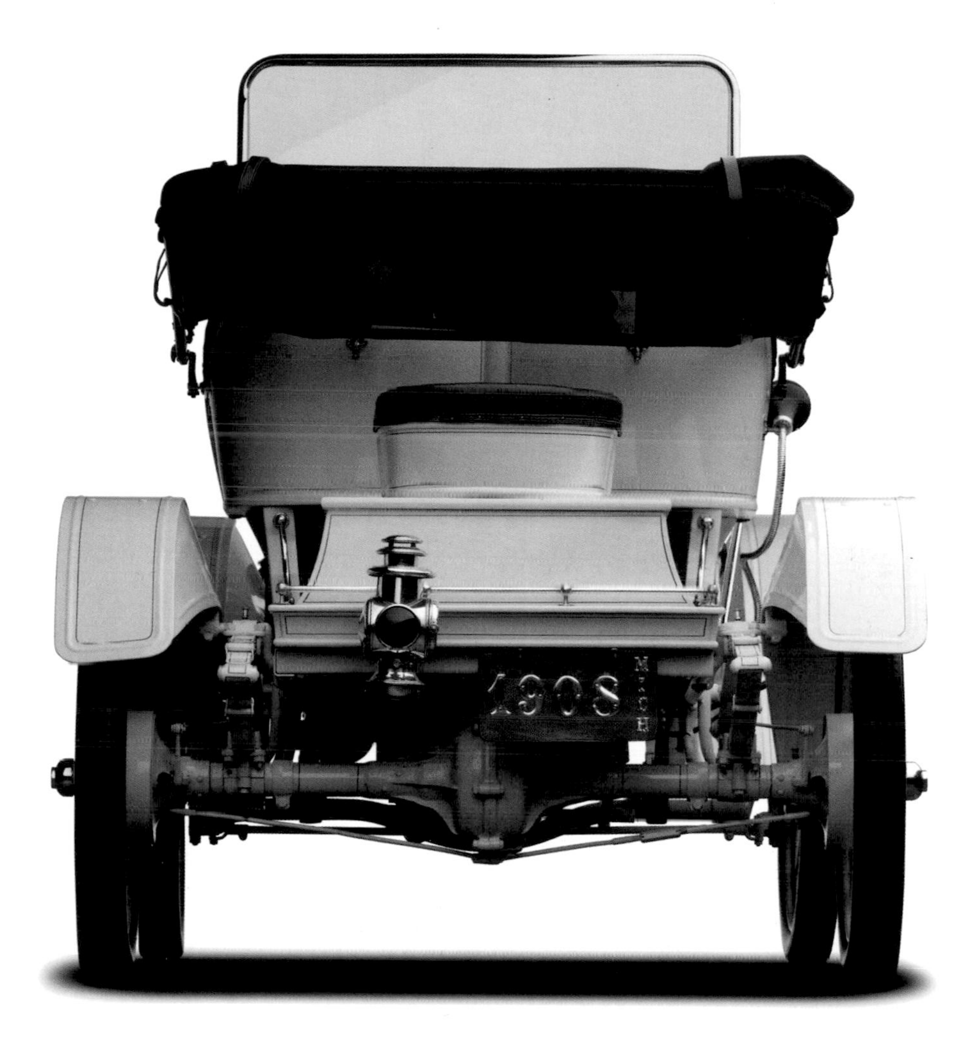

1911 OLDSMOBILE LIMITED TOURING

When Oldsmobile rolled out its Limited in 1910, luxury was clearly on the mind of William J. Mead, the former Buick executive picked by the head of General Motors, William C. Durant, to lead Oldsmobile. In fact, the entire Oldsmobile line went upscale that year, with the Limited at the top of the scale. The firm's catalog for 1910 noted that "such a car cannot be produced rapidly, therefore a limited quantity can be built." The car was colossal, and it grew in even greater size and power in 1911, when factory prices ranged from $5000 to $7000.

The Oldsmobile Limited stretched to 198 in. (5029 mm), weighed 5160 lb (2341 kg), and was even picked to transport the country's extra-large president, William Howard Taft—it was, perhaps, one of the only cars able to do so, comfortably—when he visited Monroe, Michigan, on an official presidential stop. Assisting Taft and other Limited riders in boarding the grandiose automobile were two running boards, necessary because of the entry height. At 42 in. (1067 mm) in diameter, the wheels for the Limited were huge, a feature that Olds advertised as providing the ultimate in road handling and feasible maintenance: it was claimed that the big tires would require less changing.

Although only a few more than 400 Limiteds were produced over three years, the model remained in the public eye thanks to creative promotion. A Limited appeared in artist William Harnden Foster's painting *Setting the Pace* (1909), which depicts the Olds outsizing and outrunning New York Central's *20th Century Limited* express passenger train, implying the car's unmatched power on the road. Memorable advertisements covered the walls of dealer showrooms, including a special chromolithographic version of *Setting the Pace* that also appeared in social clubs and motoring and athletic facilities. The Limited became etched in unlimited fashion in the minds of the Brass Era car–buying public.

Oldsmobile's Limited—a grand, over-the-top offering—represented an attempt by the company to position itself as the premier maker of luxury cars.

Firestone
Firestone

Oldsmobile
Limited

Firestone

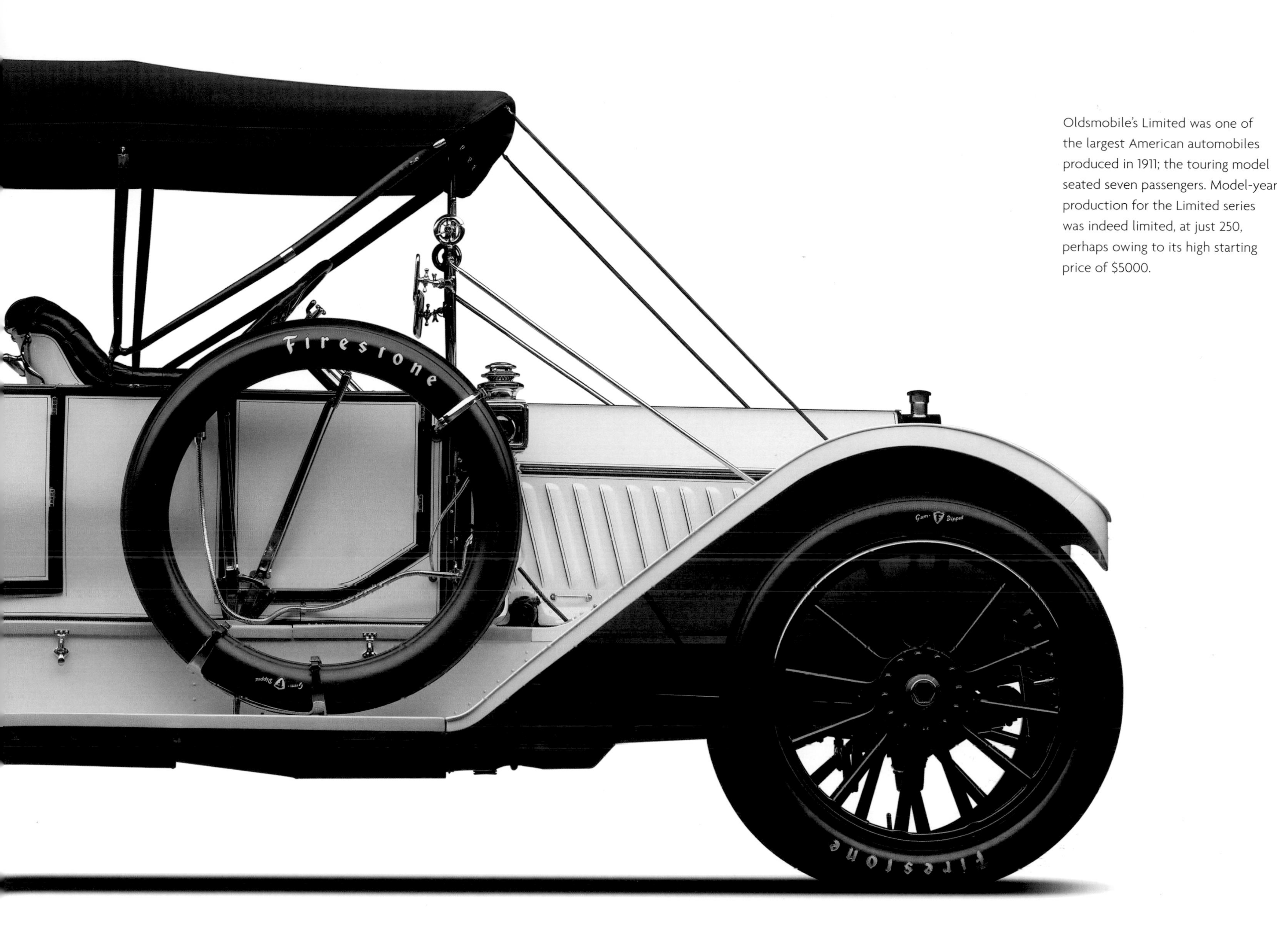

Oldsmobile's Limited was one of the largest American automobiles produced in 1911; the touring model seated seven passengers. Model-year production for the Limited series was indeed limited, at just 250, perhaps owing to its high starting price of $5000.

Ford

1914 FORD MODEL T RUNABOUT

Henry Ford's Model T was a sensation right out of the gate. Barely a year after its introduction in late 1908, the Ford company advised its dealers not to "send in any more orders until advised by this office." And demand did not let up. In addition to its low entry-level price, which purposefully allowed almost every working man to own a vehicle of his own, the "Tin Lizzie" proved both roadworthy and simple to repair. The company touted, "Drive a horse ten thousand miles day in and day out—and you'll need a new horse. The Ford will need but new tires."

In 1914, Ford drastically reduced its chassis production time, allowing for the construction of more than 300,000 Model Ts that year.

Ford addressed the growing demand for the world's bestselling automobile by developing the concept of mass production in relation to a moving assembly line. The system began in August 1913 at Ford's Highland Park facility in Dearborn, Michigan. Henry Ford realized that, in order to build thousands—not just hundreds—of automobiles a day, the assembly line needed to move. In 1914, Ford churned out more than 300,000 Model Ts; by comparison, the total production figure for the other American automakers combined was about a hundred thousand fewer.

Henry Ford's insistence on minimal—if any—year-to-year changes to the Model T was the rule of law, and the models of 1914 reflected this attitude most obviously in their uniform color scheme. Ford's famous "any color so long as it's black" policy owed its introduction to the quickening production line, which dictated fast drying times for paint. Black Japan enamel was the only paint available to Ford at the time that could perform to the new fast pace.

More than 35,000 Model T Runabouts were produced for 1914. By the start of that year, the Ford company had established a well-disciplined network of 7000 dealers, all of whom were productive in their role of order execution and delivery; they also benefitted financially from the fast inventory turnover. Factory employees, too, had something to smile about in 1914, when the work day was decreased from ten to eight hours and the hourly wage was nearly doubled.

Ford
MADE IN
U.S.A.

Opposite: The 1914 Model T had wooden artillery wheels and pneumatic "clincher" tires. Steel welded-spoke wheels became available in 1926.

Right: Although the Model T was made affordable to the masses, some niceties were retained, such as brass trim on the radiator, headlamps, and taillights.

1922 PAIGE 6-66 DAYTONA SPEEDSTER

Harry M. Jewett, who had made his fortune in coal by the turn of the century, founded the Paige-Detroit Motor Car Company in 1909, producing first of all "Paige-Detroit" automobiles, and then, from 1911, models named simply "Paige." Although the firm's slogan, "The Most Beautiful Car in America," was overconfident, its line of open sporting models was graceful and, by 1920, known for reliable speed. In 1921, Jewett proved his cars' worth by setting a land-speed record at Daytona Beach, Florida, the sandy stretch of coastline used extensively in those days for reaching unreachable speeds. Driver Ralph Mulford, who had been recruited to assist in the development of that year's Paige models, sped a stripped-down 6-66 Paige to 102.8 mph (165.5 km/h). Ever the smart businessman, Jewett pounced on the opportunity to promote his product further, producing the most famous Paige to date in 1922, the 6-66 Daytona. The tagline for the Paige Daytona was "The World's Fastest Car." Such was the power of association used.

The Paige 6-66 Daytona Speedster was one of the sleekest cars of its day—and came with plenty of punch to boot.

The Daytona was a three-seat roadster with a 366-cu.-in. engine. The third seat pulled out from the side of the car over the near-side running board, daring the adventurous to literally ride in the open. The roadster, which could do 80 mph (129 km/h) wearing full body equipment, rode on a 131-in. (3327-mm) wheelbase and was powered by a Continental engine with a bore and stroke of 3¾ × 5 in. (95 × 127 mm). Impressively, the block and cylinder head were constructed from iron, and the crankcase from lightweight aluminum. Equally impressive were the three-speed Warner non-synchromesh gearboxes. Among the finest of their day, they were renowned for their ease of use, reliability, and quietness. Indeed, Paige had clutch and transmission building down to a fine art.

It is estimated that only fifty Paige Daytonas were produced, perhaps owing to the car's high price. The Paige catalog for 1922 listed the Daytona for $2495, and it was discontinued after its second year on the market.

Opposite: The dashing three-passenger speedster came with a pull-out drawer for the third seat. It was for only the brave at heart.

Below: The 1922 6-66 Daytona Speedster was the direct result of Paige's success the year before at speed trials held on Daytona Beach, Florida.

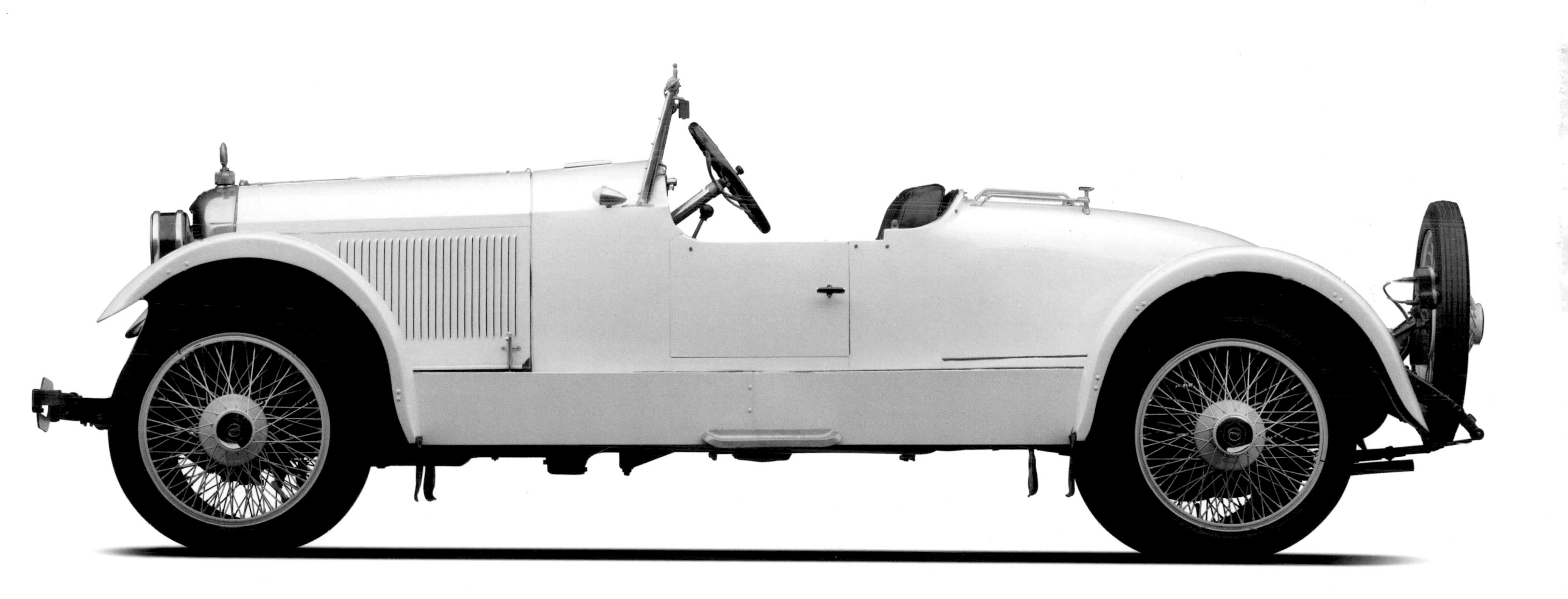

1926 KISSEL 8-75 SPEEDSTER

Times were good for the Kissel Motor Car Company in 1925 and 1926. Factory lines were running at full capacity to keep up with sales that exceeded those of any year since George and Will Kissel had formed the company in 1907. The Kissel brothers' deliberate and careful approach to car building established a reputation for quality construction that grew with the firm until its demise during the Great Depression.

Refinements for 1926 included rubber shock insulators instead of spring shackles, which, together with a new rubber barrier cushioning the engine mounts, gave the models of that year a less noisy and more comfortable ride. What attracted buyers, however, were the cars' silhouettes, especially in the case of the 8-75 Speedster. Benefitting from a chassis with a low center of gravity used by other Kissels, the 8-75 sported road-hugging and powerful lines, which were broken only by dual side-mounted spare tires.

The Kissel company gradually improved upon the "modern" design approach that it had begun in mid-1917. By 1926, Kissel was wooing movie stars and other high-profilers with factory-produced "custom look" cars. By this time, interest in straight eights was reaching a crescendo. In speedster form, the 8-75 could give the era's fast Lincolns a run for their money, reaching 85 mph (137 km/h) with the in-line, eight-cylinder, 310-cu.-in. engine—rated at 71 hp at 3000 rpm—used across the model range. From 1924, the engine was based on a Lycoming block fitted with Palmer-designed cylinder heads and aluminum oil pan, pistons, and connecting rods. The Kissel 8-75 Speedsters' double-drop frames and Werner-designed bodies gave them a look to match their performance, while also living up to the company's reputation for advanced design and dependability.

Kissels of the mid-1920s were handsomely designed and featured custom-body looks. Such styling was most noticeable on the 8-75 Speedsters.

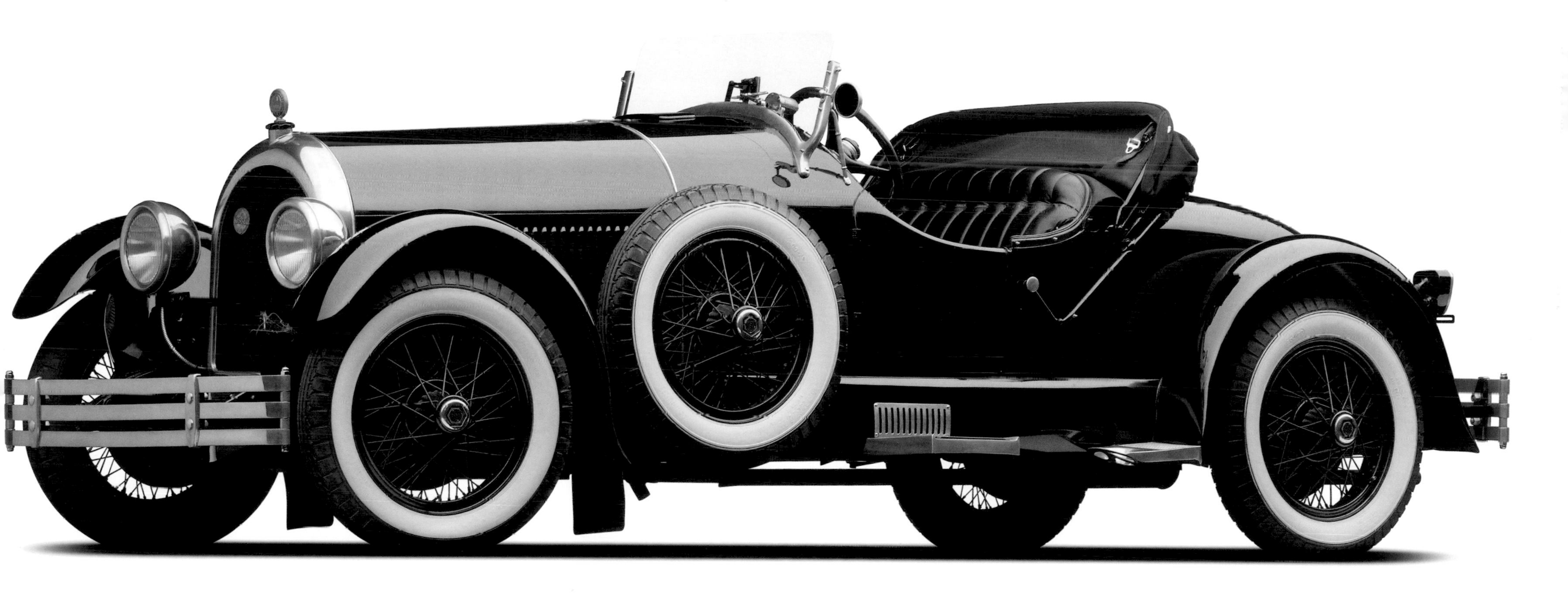

In 1926, Kissel bragged that it had pioneered more new automobile styles than any other American manufacturer. Of the 35,000 cars produced by the company during its lifetime, only 150 are known to exist today; just four of those are 8-75 Speedsters.

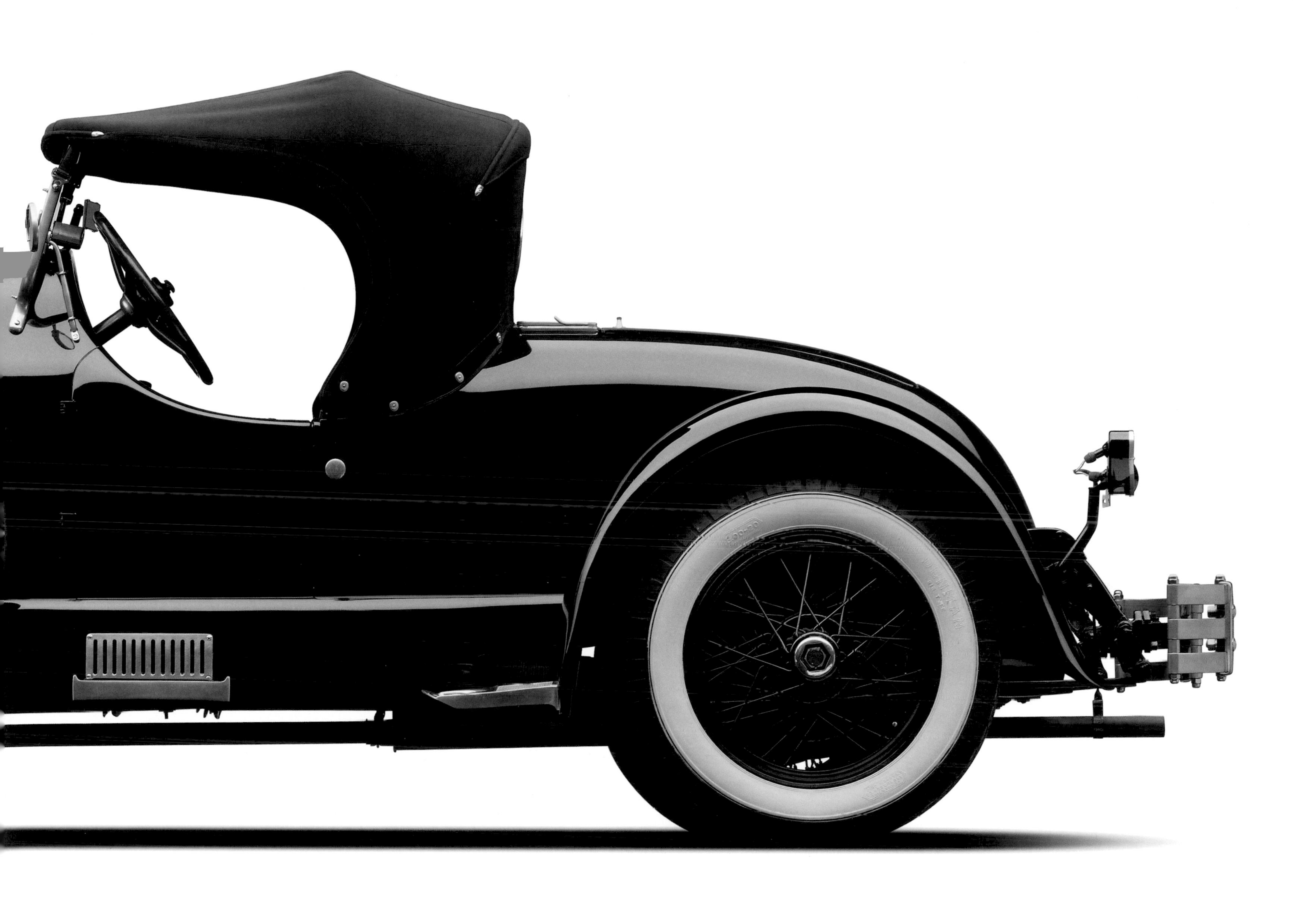

Left: In 1926–27, the 8-75 Speedster was often referred to as the sportiest car on the market. Looking at the car head-on, it is hard to disagree.

Opposite: Engines in the Kissel 8-75 model range were based on a Lycoming block. The aluminum cylinder heads and oil pan were cast in Kissel's own foundry.

KISSEL
No 3527 MODEL 75

1926 PACKARD 326 ROADSTER

The "three Ps" (Packard, Peerless, and Pierce-Arrow) put the sparkle into the U.S. auto industry's Gilded Age. Comfortably atop the industry's luxury summit in the 1920s, Packard introduced its third series of Packard Sixes in February 1925, and they were produced through August 1926 on a 126-in. (3200-mm) wheelbase. The series was promoted in a range of confident advertisements, which featured such tag lines as "When You Arrive in a Packard," "Born in the Lap of Luxury," and other high-mannered phrases. At $2785, the 60-hp Model 326 Roadster was more reasonably priced than some of the earlier Packards, but was still beyond the reach of most car buyers.

Packard's powerplant for its "Third Series" Sixes increased in horsepower in 1926, adding to the marque's prestige.

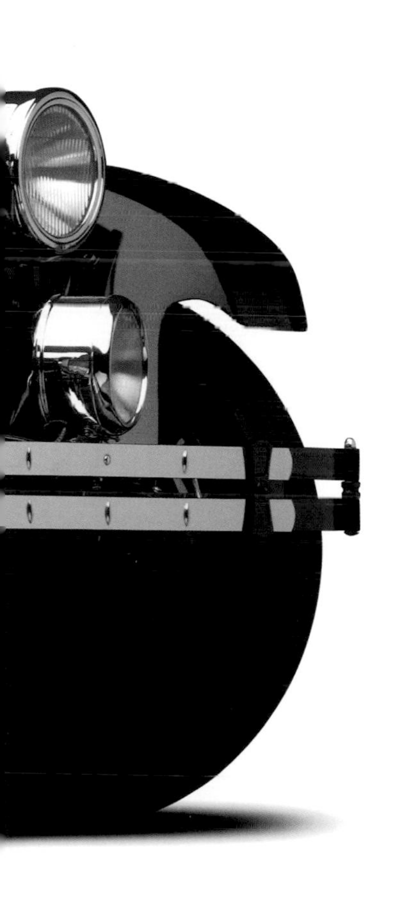

The Single Six, as this sporty roadster was categorized in the Packard catalog for 1926, had an L-head six-cylinder engine that produced 60 hp at 3200 rpm. Revisions for the Six lineup of 1925–26 included a simplification of the brakes in late 1925 with the adoption of the Bendix braking system. Mechanical brakes were now available all around. A new paint finish was also used: a DuPont development called Duco, a sprayed-on pyroxylin lacquer that dried within a few hours. The change from using varnish finishes, which could oxidize or dull, was an illustrious one: the new lacquer added durability as well as outstanding color. In early 1926, two-tone combinations became commonplace for open cars, including the sporting variety.

What made Packards of the era so special was their singular design. The Packard body engineer in charge at the time was Archer Knapp, who said, "What is needed is not an artist, but a real body designer . . . as it is necessary nowadays to design a vehicle as a whole, in order to have harmony of lines in the entire car." This initiative worked well for Packard: the 326 was extremely popular. Model-year production for the 326 lineup was 24,668, helping Packard break overall company records in sales, production, and earnings the year it was brought to market.

Left: In addition to disc wheels, the 326 was available with a wider selection of colors as standard. It also sported a newly developed lacquer finish.

Right: The 326 was unashamedly promoted as a luxury vehicle. The inclusion of golf-bag doors just forward of the rear wheels gives an indication of the company's target audience.

Left: The 326 came equipped with a rumble seat, a popular option of the classic era that allowed one or two passengers to ride in open-air style behind the main cabin.

Opposite: As if to emphasize its luxury credentials, the Packard was fitted with Pilot-Ray driving lights, first seen on only the most expensive cars. The lights turn in conjunction with the tires.

Packard

1927 LASALLE 303 ROADSTER

For a car-crazy public, most Americans in the 1920s did not know what they were missing. That is, until 1927, when Cadillac came out with a sporty "companion" car called the LaSalle. Custom bodywork fulfilled the desire for expression and distinction only to a point—and only for those who could afford it. The vast majority of models filling America's streets were boxy and relatively nondescript. That all changed when Fred Fisher and Alfred Sloan of General Motors tasked Californian Harley Earl with designing a car that would fill the price gap between the highest-priced Buick and the lowest-priced Cadillac. The resulting LaSalle, the first production car designed by a stylist, achieved mass appeal.

The LaSalle was sold as a "companion" marque to Cadillac from 1927 to 1940, ushering in a new era of American styling.

Earl was acquainted with the sleek lines on some of the European marques of the day, automobile styling that provided a lightness and flair never seen on U.S. cars. With Hispano-Suiza on his mind (Earl had recently seen the luxury firm's exhibit at an auto salon in Paris), he set out to design an American automobile as low and as long as possible, given the project's parameters. Not surprisingly, the new Model 303 emerged with a European marque-like narrow radiator, sweeping clamshell fenders, and unusual two-tone paint effects.

Initially, Earl worked on the LaSalle as a consultant to Cadillac, but the design was so successful that General Motors hired him as head of its new Art and Color Section.

Built by Cadillac and to Cadillac standards, the LaSalle soon emerged as a trend setting automobile within GM. In addition to a fresh styling direction on the exterior, the LaSalle also had a newly designed engine. Based on the standard Cadillac V-8, the new powerplant had a more economical design, despite its cylinder heads being "ribbed" for aesthetics. In the end, it proved superior to the original in every way.

Built by Cadillac to its usual high standards, the LaSalle soon emerged as a trend-setting automobile. The roadster was available in two-tone color combinations, at a time when dark colors remained the norm.

Opposite: A cast figurine of René Robert Cavelier, Sieur de La Salle, was used as a hood ornament on early models. La Salle, from whom the marque took its name, was a seventeenth-century French explorer remembered for his travels along the Mississippi River.

Right: The LaSalle was offered in a full range of body styles, including custom designs built by Fisher and Fleetwood Metal Body. For the wider industry, it officially heralded the beginning of the Harley Earl era of design influence.

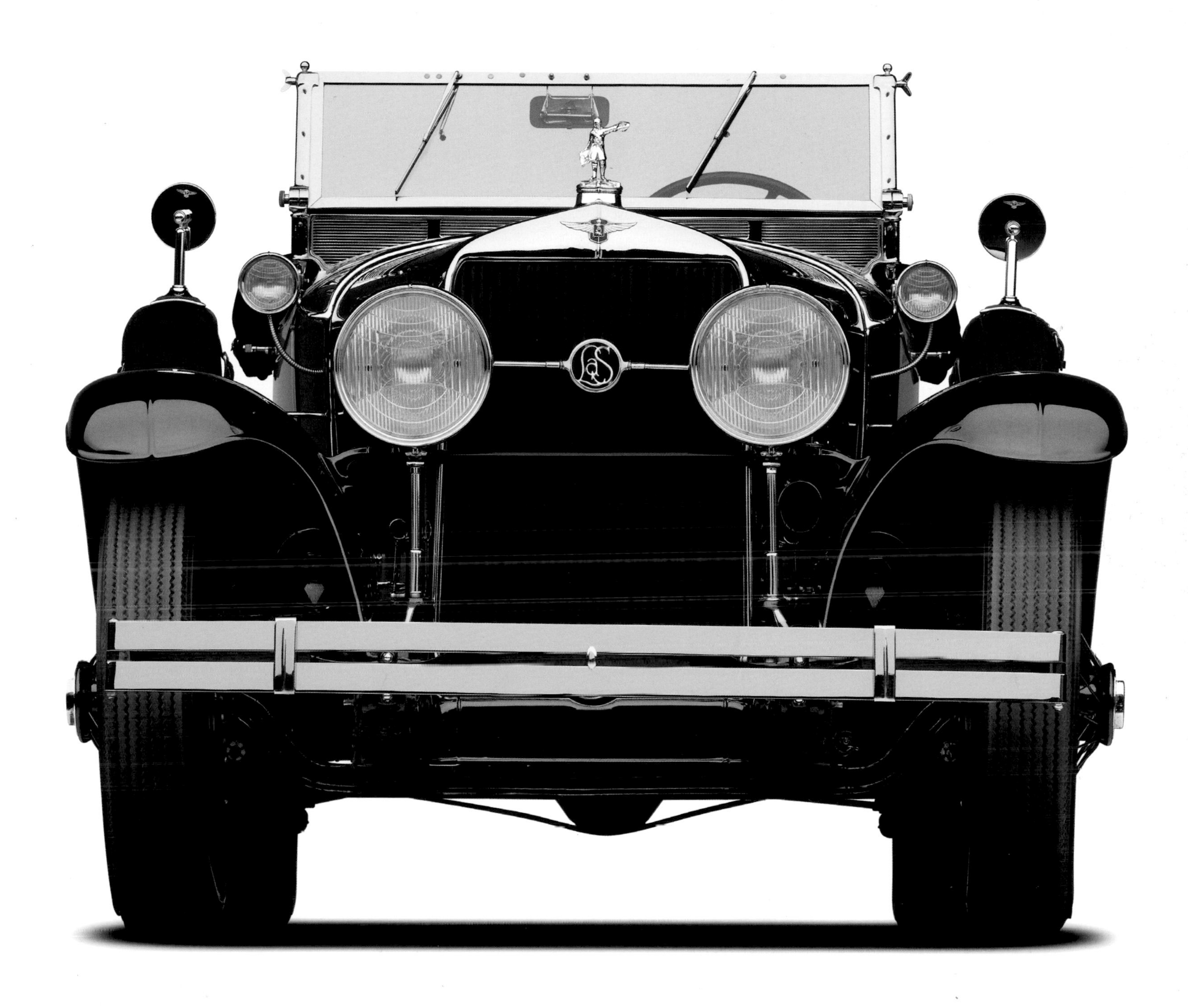

1929 DUPONT MODEL G SPEEDSTER

As a limited-production luxury marque, DuPont produced only 537 cars during its lifetime, from 1919 to 1931. The DuPont automobile was the child of E. Paul du Pont, who set up shop in Wilmington, Delaware.

The introduction in 1928 of the Model G was in part an answer to Duesenberg's just-released Model J, a powerful entry in the luxury segment, with track speeds to prove it. DuPont's Model G was powered by a Continental side-valve eight-cylinder engine with a 323-cu.-in. displacement, which produced 125 hp at 3200 rpm.

A short-wheelbase adaptation of the Model G was made in 1929: the speedster, a two-seat production copy of a four-person Le Mans car entered in that year's contest. DuPont intended to race the car in the Le Mans 24 Hours, to compete with archrival Stutz. DuPont's entry was refused, however, because Le Mans was not open to two-seaters. Quickly, DuPont built two cars that did meet regulations; only one was entered, and it retired early in the race with gearbox issues. The production speedsters were guaranteed by the factory to exceed 100 mph (161 km/h).

The first Model G Speedster—which, with a 135-in. (3429-mm) wheelbase, was 6 in. (152 mm) shorter than its huge siblings—was seen at the New York Automobile Show of 1929. Its rounded grillwork was framed on either side by Woodlite "cat's eye" headlights. Flanking the headlights were elegantly curved, one-piece fenders that swept from the front of the car to the rear. The show car was the object of love at first sight: actress Mary Pickford bought it as a birthday gift for her husband, Douglas Fairbanks.

The Model G was the most successful DuPont to date, with production figures reaching 273 in total. Bodies were coached by Merrimac, Waterhouse, and Derham.

Although DuPont produced relatively few automobiles during its existence, the company was renowned for the luxury, quality, and style of its vehicles.

The Model G used a conventional ladder-frame chassis with live axles on both ends and drum brakes on all four wheels. The car's solid mechanicals and styling led to it becoming one of DuPont's most popular offerings.

1930 DUESENBERG MODEL J ROADSTER

Fred S. Duesenberg made his name in powerful engines and racing before E. L. Cord came on the scene in the 1920s, purchasing Indianapolis-based Duesenberg as part of his growing empire. (Cord's businesses included Duesenberg, Cord, Auburn, and other transportation concerns.) The Model J was a product of both men, however, and was introduced on December 1, 1928. Fred Duesenberg was a perfectionist engineer, and he joined Cord in promising that the new Model J would be "the world's finest motor car." They were certainly on target.

The Model J, with a price tag of $8500—sans coachwork; that was $3500 or more extra—featured a ground-shaking straight-eight engine, impressive at 4 ft (1.2 m) long. The engine block itself was a marvel, with chrome and nickel studding the traditional apple-green enamel, all under the shadow of a giant cylinder head with thirty-two valves and twin overhead cams. The engine shared the rare use of aluminum alloy throughout. Compared to other automobiles of the era, which tended to use cast iron, the Model J gleamed with aluminum from stem to stern.

Un-supercharged, the J produced nearly 265 hp and recorded a maximum speed of 116 mph (187 km/h); it was capable of 94 mph (151 km/h) in second gear. In terms of power, the Model J was top of the food chain, putting out more than three times that of its competitors. Indeed, it was the fastest automobile on the market. The roadster was available on a 141 3/4-in. (3600-mm) wheel-base, 12 in. (305 mm) shorter than that of other body styles.

Only high society was able to afford the hefty price tag of the J. Approximately 470–480 Model Js were produced in total, including such later variations as the SJ, JN, and SSJ. The sales target set for the Model J's first year of production was 500, but that was missed by approximately 200, thanks largely to the Depression.

The Duesenberg Model J was celebrated as the ultimate American automobile. Its significant price tag put it within the reach of only the wealthiest members of society.

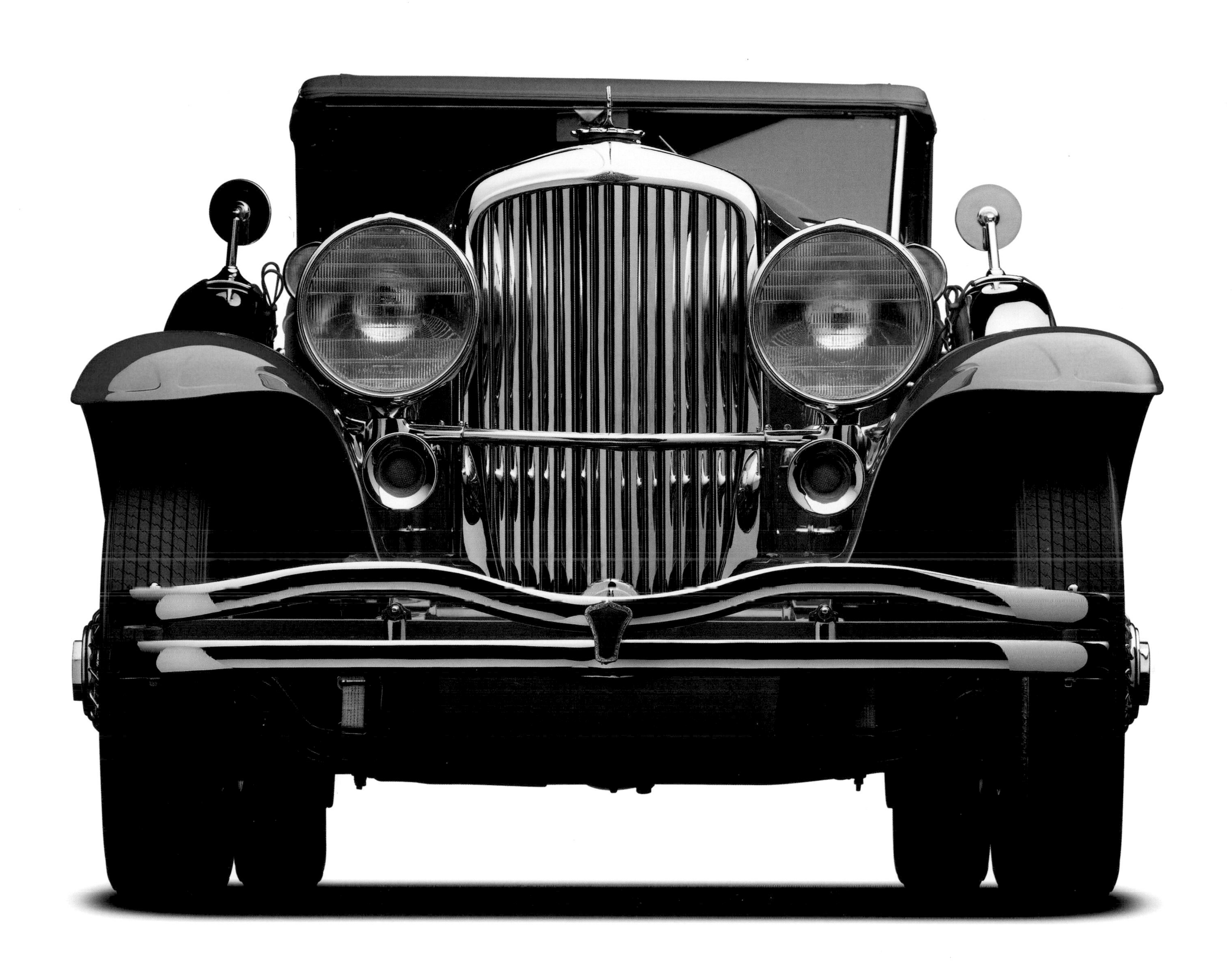

Fred Duesenberg's engineering skills were seen at their best on the groundbreaking Model J. When the Depression hit in October 1929, only some 200 cars had been built. An additional 100 orders were filled in 1930.

1930 FRANKLIN SERIES 147 RUNABOUT

Beginning in 1924, Franklins were built according to a more conventional mindset, including the use of a "false" radiator in the body design. From the time they first hit American roads in 1902 until the last car left the factory in 1934, all Franklins were air-cooled, thus having no need of a traditional water-cooled radiator. From the time the false radiator was included, Franklin employed a greater variety of body styles and, eventually, a number of coachbuilding firms.

Franklins were conservatively styled by nature: it was the predominant philosophy of John Wilkinson, Franklin's vice-president. In close cooperation with Franklin Automobile Company founder Herbert H. Franklin, Wilkinson was in charge of the company's namesake automobiles from the start. It was said that the design-engineer-in-charge had absolutely no patience with the whims of sales people. Fortunately for the sake of corporate peace and stability, the sales force nodded in approval of Wilkinson's direction. His influence on quality engineering remained after he left the company in the mid-1920s.

By 1930, Franklins had gained weight and grown larger. But the bodies were sexier than ever. Raymond Dietrich was now in place as Franklin's chief designer, and with the help of Erwin "Cannon Ball" Baker's endurance exploits, Franklins sold well until Wall Street crashed. The press went wild over Franklin's new six-cylinder engine, which debuted in December 1929 for the following year's models. Newspaper accounts applauded by calling it the greatest advance in air-cooling history. Also notable were the engine's larger valves, intake manifold, and carburetor, complemented by a smaller and more efficient fan. An engine capacity of 274 cu. in. resulted in an increase in horsepower from 65 to a maximum of 95 at 3200 rpm. The Series 147 made reaching 80 mph (129 km/h) an easy feat, with *The Motor* noting that it could compete with the speedsters of the day, including the big Lincolns and the open-touring Mercedes, Stutz, Cadillac, and Lancia units.

All Franklins used air cooling. The company considered it a simpler and more reliable system than water cooling.

Below: Powered by a new six-cylinder engine, and with one of the highest power-to-weight ratios of the period, the Series 147 could easily hold its own against the speedsters of the day.

Opposite: As recognized by this striking hood ornament, Franklin engines were also used to power numerous light planes.

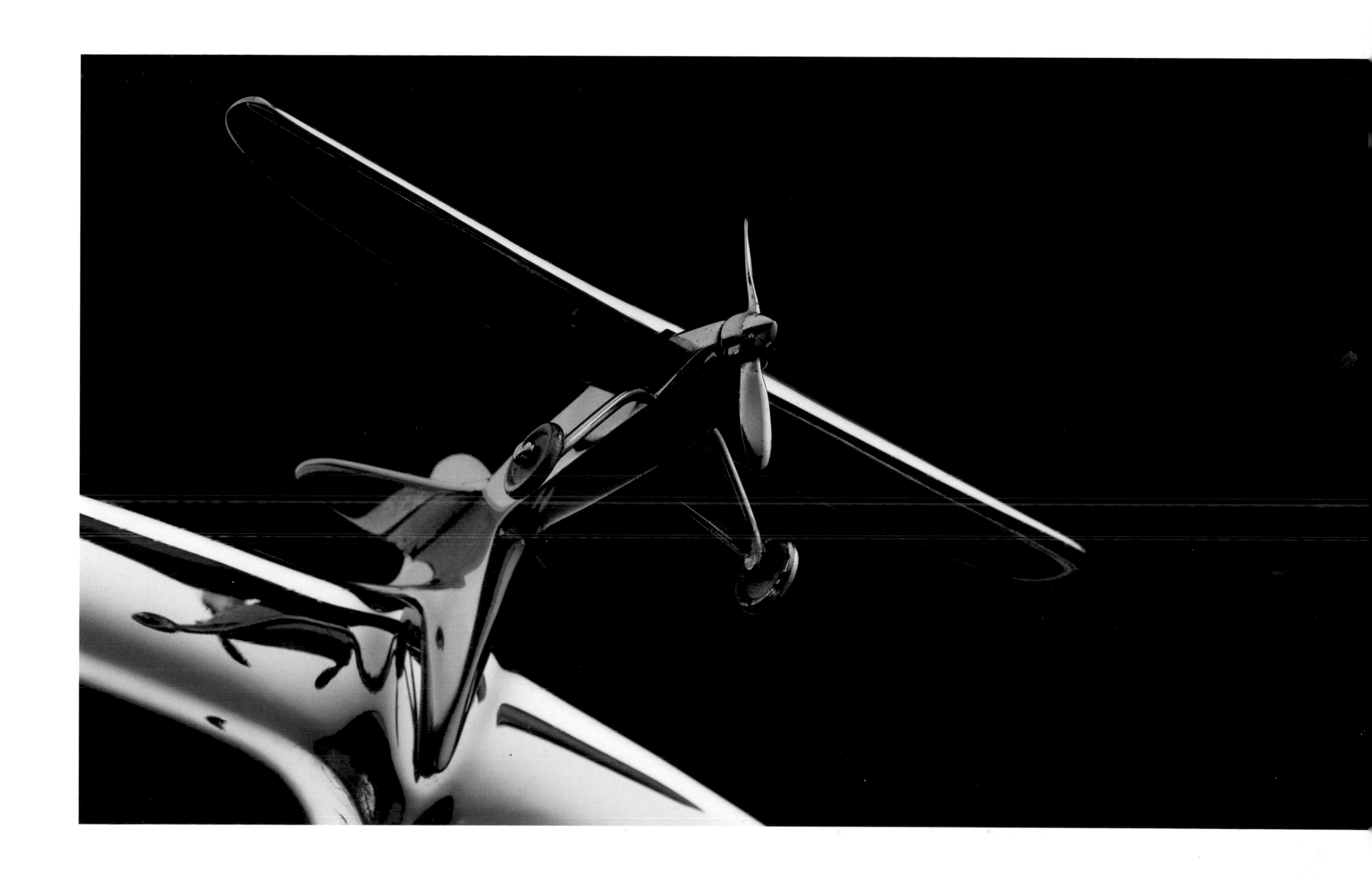

1930 STUTZ MODEL M SPEEDSTER

Harry Stutz started his business as the Stutz Auto Parts Company in Indianapolis in 1910. Stutz cars were promoted in competitions from the start, and would eventually compete in the Indianapolis 500, the Vanderbilt Cup, and other endurance races; in 1928, Stutz took second place at Le Mans. But it was with the introduction in 1926 of the Vertical Eight engine—so named to distinguish it from Packard's Straight Eight—that Stutz could lay claim to being a fully fledged luxury marque. A 92-hp straight-eight engine was kept current with a revision to 115 hp in 1928. Longer models, with a 145-in. (3683-mm) wheelbase, were added, and soon all were available with series-custom coachwork.

The Model M's low profile, made possible by Stutz's sporty underslung drive design, was widely impersonated by other automakers in the 1930s.

Stutz's Vertical Eight was of advanced design, with an overhead camshaft, dual ignition, nine main bearings, and full force-feed lubrication. The influential power unit pioneered a silent, reliable, self-adjusting chain drive on an overhead-cam engine—quite an achievement in the era of noisy bevel gears.

The public welcomed the swift Stutz speedsters with open arms, no doubt with memories of the beloved Bearcat fresh in their minds. Although the underslung worm drive used by the Model M was eventually phased out across the industry, the car's resultant low-body design sparked the trend for automobiles with a height of 20 in. (508 mm) or less. Other automakers' models were seen sporting very similar designs, including those of Stearns-Knight, Pierce-Arrow, and Nash.

By 1930, fatal wounds had slowed the company's progress. Lawsuits streamed in during the late 1920s, and success on the track was sporadic at best. The financial strain took its toll, and fewer than 1500 cars were produced for the whole of 1930. Sales declined even further in the following years, and in 1934, after the production of only six Stutz cars, the factory closed its doors.

The Stutz Model M Speedster of 1930 remains one of the company's most handsome models, a testament to what had made Stutz great.

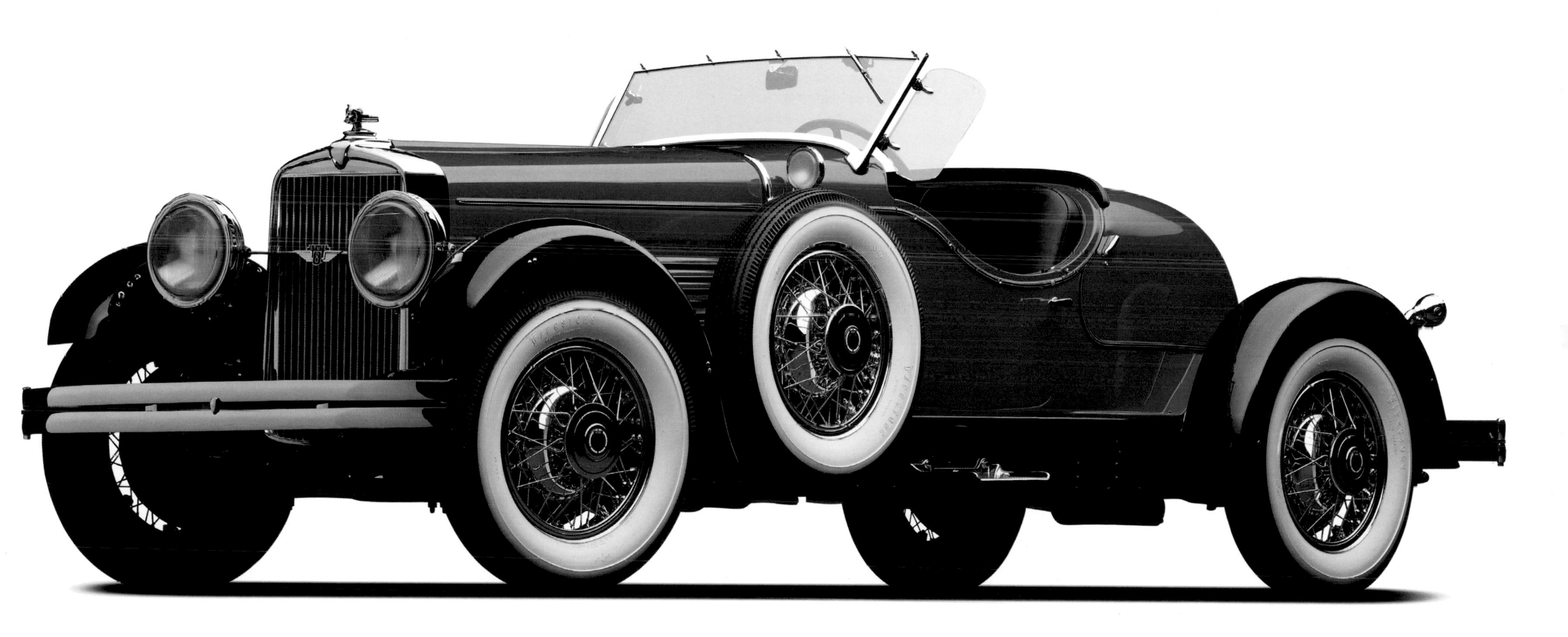

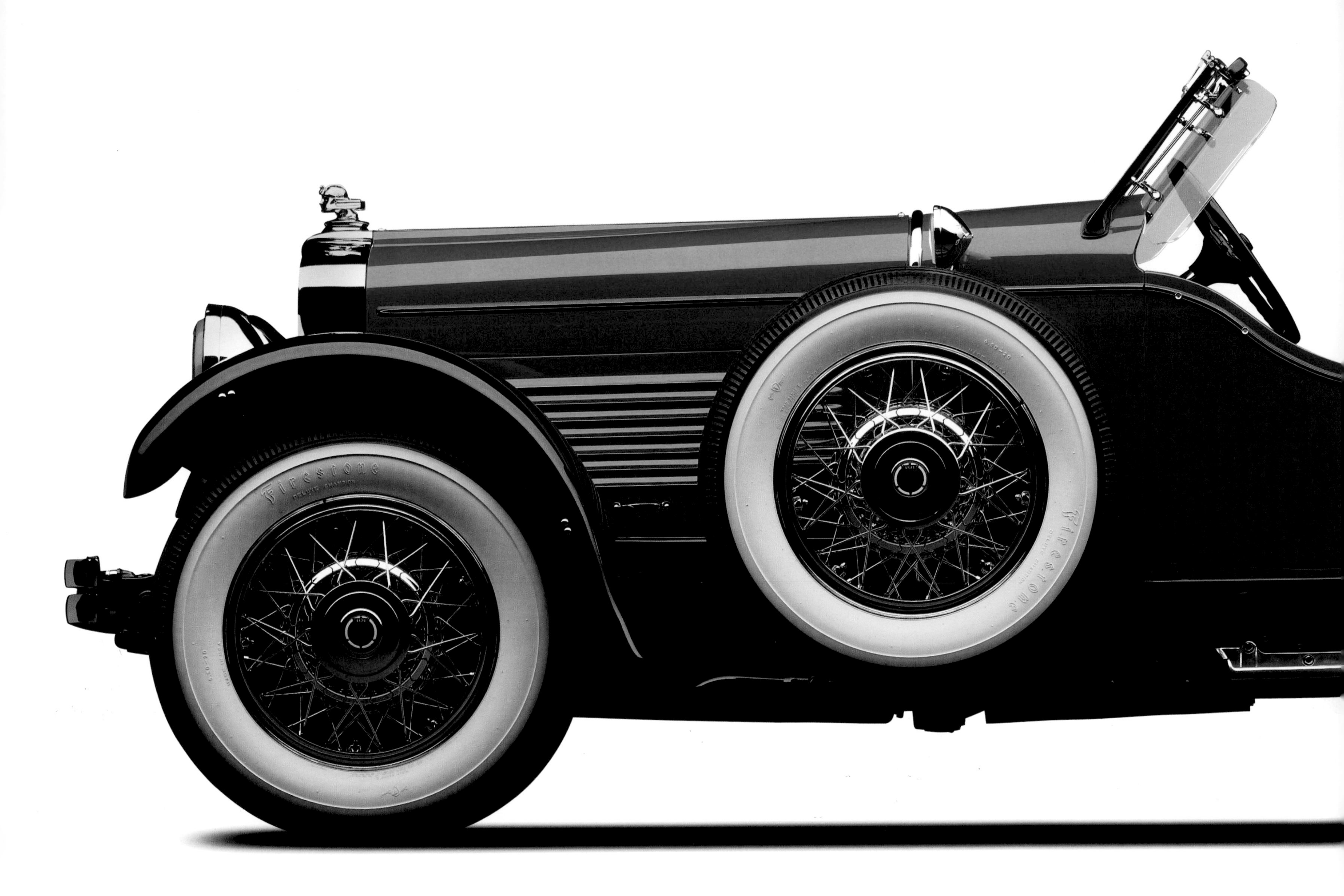
Firestone
DELUXE CHAMPION

The Stutz Model M Speedster was one of the sportiest cars of its day. The two-seater was priced at $3175.

Below: Stutz's Vertical Eight was well received by the industry. In the first week of the engine's availability, Stutz received orders worth $3 million.

Opposite: Luxurious style was matched by Stutz's history of success in racing and endurance events.

CHAMPION
STUTZ

1931 CHRYSLER IMPERIAL ROADSTER

Walter P. Chrysler was the consummate "ideas man." Endeared with equal and copious amounts of mechanical-engineering prowess and design acumen, he rolled out the first car bearing his name in 1924, the Chrysler Six. From that first model, his marque became synonymous with quality. To compete directly with such luxury heavy-hitters as Stutz and Marmon, Chrysler came out with a new model, the Imperial, in 1926. Following its rookie year, the newcomer sold 2000 or more units annually.

Introduced in 1926, the Chrysler Imperial was the company's top-of-the-range model. Highlights of the car's redesign in 1931 included sweeping fenders and wire wheels, which remained a standard wheel treatment until the 1940s.

Beginning in the summer of 1929, Chrysler upped the ante of his company's Imperial offerings by matching Packard's and Pierce-Arrow's engine capacity at 384³/₄ cu. in. The engine-size increase created considerable horsepower increases as well, with 125 hp in the Silver Dome–engined models, and 135 hp in the high-compression Red Head–engined versions. By the time the Great Depression hit, the Imperial was firmly entrenched as Chrysler's flagship automobile. Priced at the top of the scale for the firm's cataloged cars, a custom-bodied Imperial Roadster by LeBaron cost $3295, over $2000 more than the price of a factory-built seven-passenger sedan.

In 1931, the Imperial underwent a redesign, with Chrysler wrapping its new straight-eight engine in bodywork of classic beauty; in marketing materials, the car was referred to as the "Imperial Eight." The totally new Imperial sported a V-type radiator, sweeping fenders, and an angled windshield that presented frontal features similar to those of the groundbreaking Cord L-29. Hydraulic shocks and adjustable front seats were just two of the attractions of this grand model, and a variety of optional custom bodies were made available, including individualized craftsmanship from coachbuilders Waterhouse, Derham, Locke, and Murphy. Semi-custom bodies were available from LeBaron.

In mid-1931, Chrysler's Imperial CG series featured dual sloping windshields, new vertical hood louvers, and carryover interior sun visors. A custom convertible was offered the following year for a wallet-busting $3995.

Below: Chrysler positioned its Imperial as a prestige model to rival the offerings of Cadillac and Lincoln. The majestic presence of the redesigned 1931 Imperial spoke of pure quality.

Opposite: A V-type radiator was just one of several new features found on the 1931 Imperial. The introduction that year of a new straight-eight engine was matched by long hoods and fenders, making the Imperial a true classic.

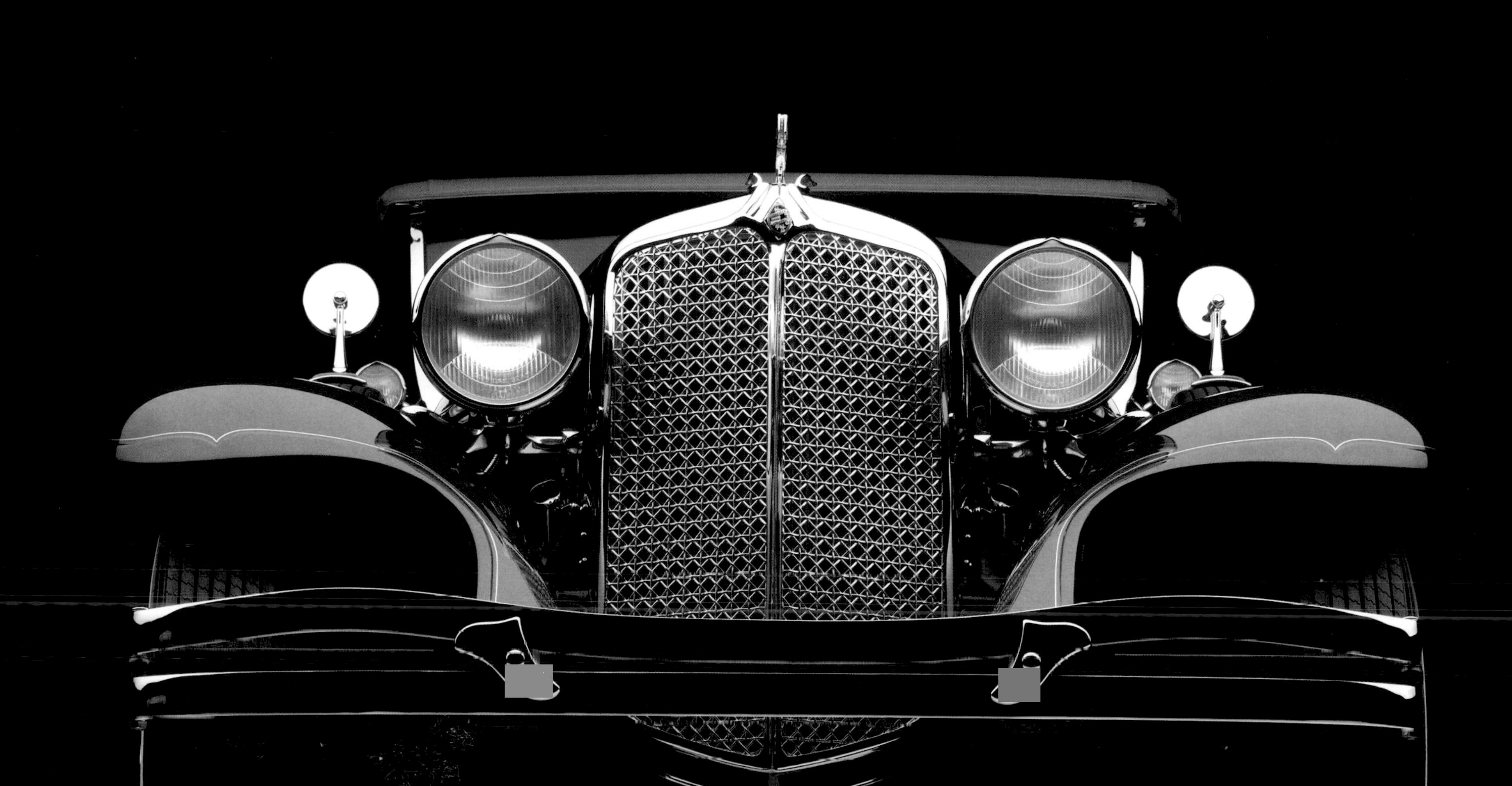

1931 PIERCE-ARROW MODEL 41 CLUB SEDAN

The Pierce-Arrow Motor Car Company was struggling financially even before the tremendous woes of the Great Depression began. In 1928, the company's six-cylinder era came to an end, and the firm was bought by Studebaker, two developments that marked the last days of the great luxury marque. In 1931, the classic cars' stature crested with Pierce-Arrow's large, straight-eight models. The chassis of that year was more popular than ever among well-to-do buyers, and custom coachwork from such builders as Brunn, LeBaron, Derham, and Willoughby was widely seen.

The Pierce-Arrow was not just a car. It was also a status symbol, favored by movie stars, corporate tycoons, and royalty.

At the start of the 1930s, the factory in Buffalo, New York, was producing standard body types with more attractive lines than previous years, particularly the Model 41 "Salon Group" series, which was described by the company catalog as "Pierce-Arrow at its best." These cars—the largest available, with a 147-in. (3734-mm) wheelbase—were also offered customized by the specialist coachbuilders. Transmission upgrades took place in 1931, most notably in the form of a new free-wheeling three-speed as standard. Acting in second and high gear, the free wheel could be locked out if desired, allowing the transmission to "coast" at cruising speeds, which permitted the engine to idle most of the time. The result was a much quieter engine, not to mention reduced fuel costs.

The attention to detail in Pierce production included fifty-five stages of body finishing that required seven separate inspections. Fourteen coats of lacquer were applied to specially selected, high-quality straight-grained northern white ash, an expensive, superbly durable wood. This standard bodywork was completed in the company's own shops, which were comparable to those of the custom builders of the day.

The most noticeable changes to the models of 1931 included a larger radiator, wider doors, and more streamlined running-board aprons and fender lines. Inside, there was more room, although the overall effect was that of a lower, sleeker automobile, a true attribute of the luxury marque of the era.

Below: Among American makers of luxury cars, Pierce-Arrow was rivaled only by Peerless and Packard. The marque's archer hood ornament remains one of the most recognizable.

Below: Several coachbuilders were commissioned to build bodies on the Pierce-Arrow chassis of 1931, including Judkins, LeBaron, Willoughby, and Rollston. Interlacing wire wheels added to the prestigious styling.

Opposite: Nestled under the hood of the Model 41 was a straight-eight engine, which remained in use—with modifications—through 1938.

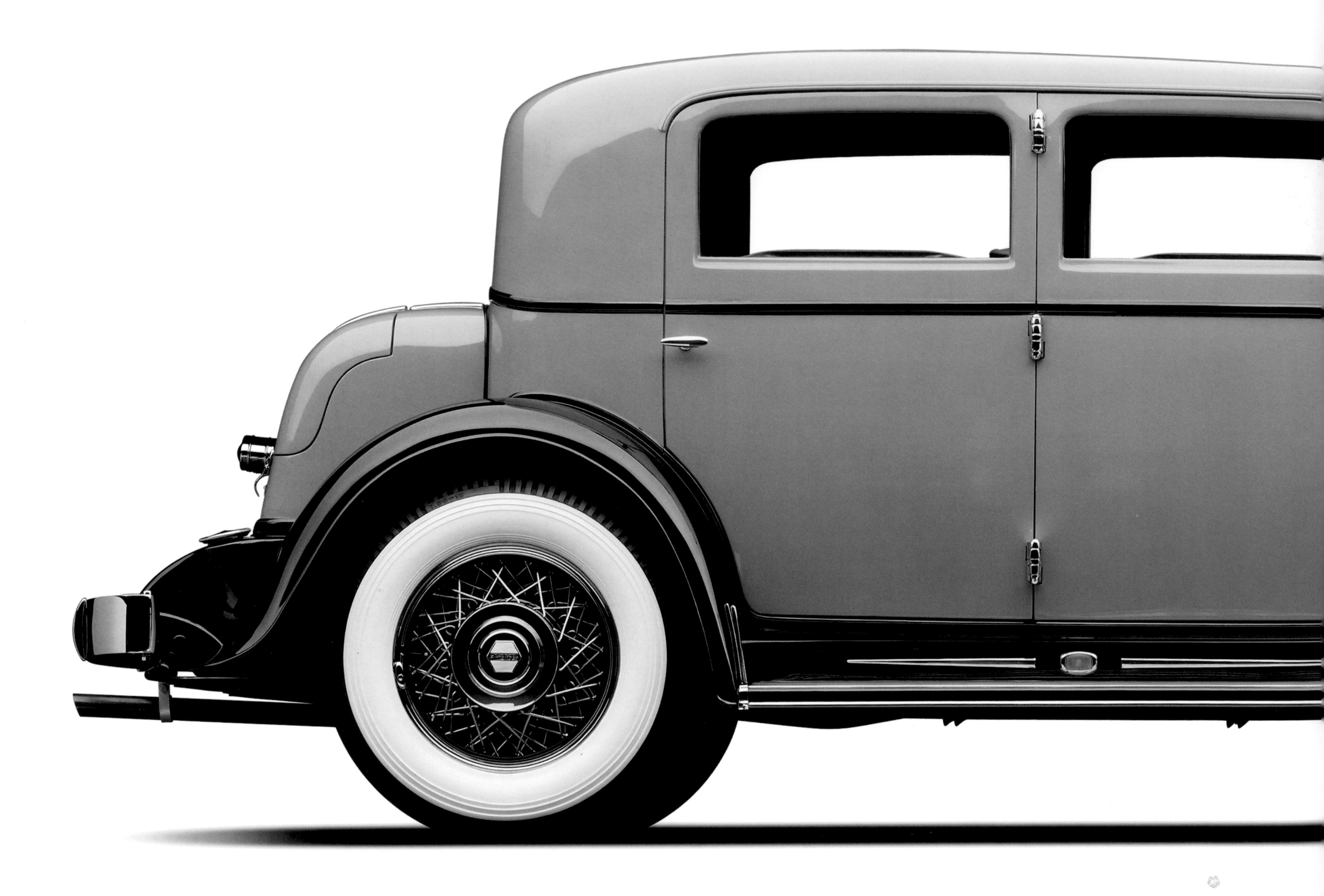

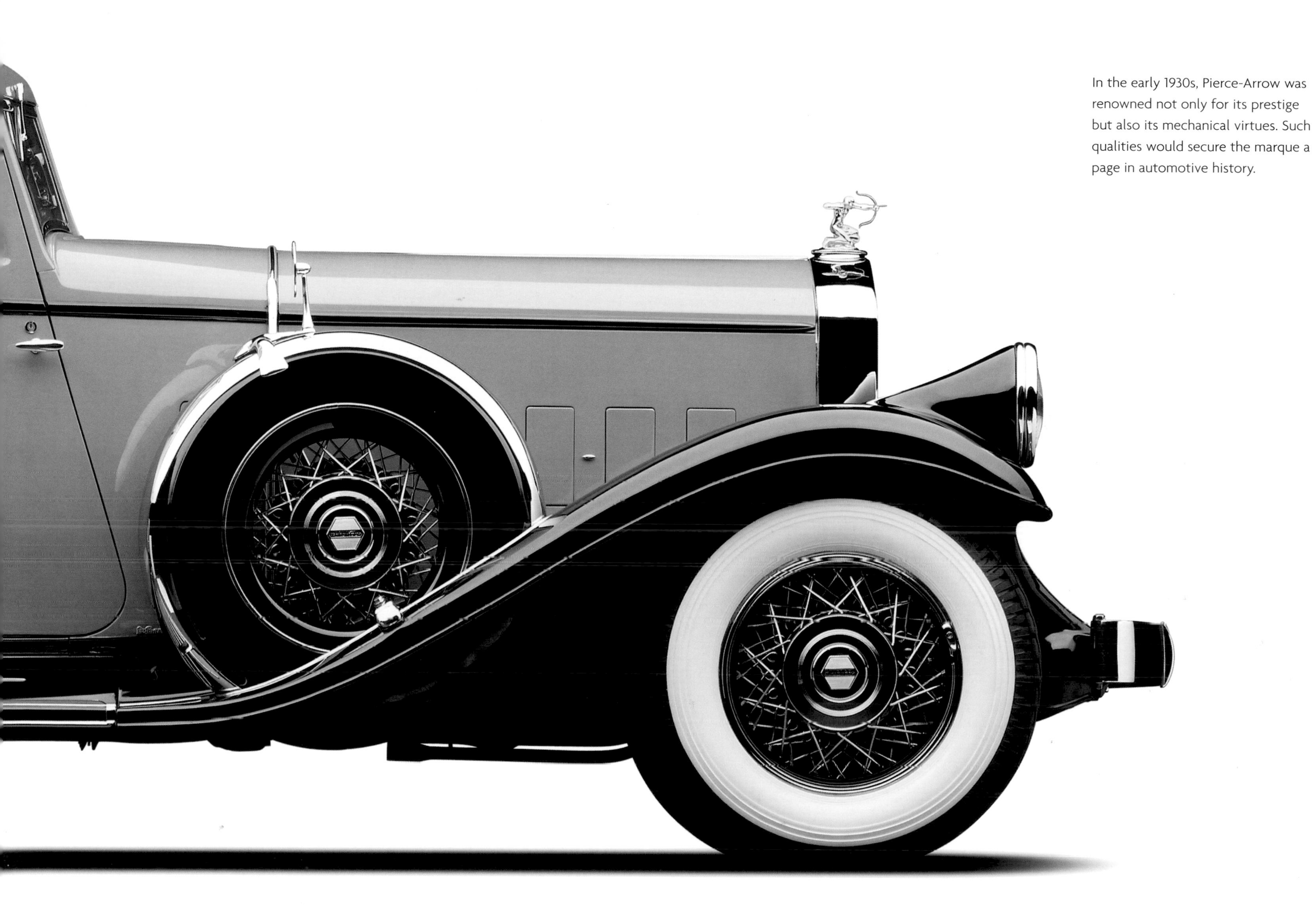

In the early 1930s, Pierce-Arrow was renowned not only for its prestige but also its mechanical virtues. Such qualities would secure the marque a page in automotive history.

1932 PACKARD TWIN SIX SPORT PHAETON

The production of luxury cars during the darkest days of the Depression may seem counterintuitive, even ludicrous. But for the high-end auto manufacturers, desperate to retain what share they had of the shrinking luxury market, offering their most glorious models made good business sense. So it was that in 1932, Americans witnessed some of the most striking creations aimed at the highest-end car buyer.

Packard Motor Car Company went to the New York Automobile Show in January 1932 with two new models, one of them the V-type twelve-cylinder Twin Six. The Twin Six's engine—a singly cast 67-degree V-type, complete with aluminum alloy pistons and a four-bearing crankshaft fitted with a vibration dampener—produced 160 hp with a bore and stroke of $3\frac{3}{8} \times 4$ in. (86×102 mm) and a displacement of $445\frac{1}{2}$ cu. in. The car itself was available in two lengths of wheelbase: 142 in. (3607 mm) and 147 in. (3734 mm).

The Twin Six cemented Packard's prominence in the luxury segment, capturing 35.6 percent of that market (number two was Cadillac, with 16.9 percent).

As Packard's first offering in the multi-cylinder race, the Twin Six became a pillar of American prestige, and featured some of the best American coachwork by such firms as Brewster, Fleetwood, Judkins, and LeBaron. Packard hired Raymond Dietrich to design its own in-house bodies. Though conservative by nature, these standard bodies were offered in nine showy designs under the "Individual Custom" label. All these designs were later touched up by Count Alexis de Sakhnoffsky to include longer front hoods that matched the width of the body.

Sport Phaeton prices ranged from $4090 for a standard model to $6300 for the Individual Custom edition. All Twin Sixes were produced as part of Packard's Ninth Series lineup, as either a 905 or a 906 model. Production totaled just 549 cars.

Packard's Ninth Series Twin Six marked the marque's entry in the multiple-cylinder power race.

One of Packard's finest "sports luxury" models to date, the 1932 Twin Six was rated at 160 hp and could reach 100 mph (161 km/h). Prices for the Sport Phaeton ranged from $4090 to $6300. All price points, however, bought a lot of car.

Left: The vertical bars of the radiator grille concealed the Twin Six's twelve-cylinder V-type, designed by Jesse Vincent. The car was later renamed the Packard Twelve, and remained in production through 1939.

Opposite: Packard's "Goddess of Speed" hood ornament adorned its cars in the early to mid-1930s. The winged goddess holds a tire in her outstretched arms.

1934 STUDEBAKER COMMANDER LAND CRUISER

Studebaker appeared ready to share the fate of a number of its competitors during the tough economic times of the early 1930s. In March 1933, the company went into receivership. But a fortunate series of events kept Studebaker in business. The firm managed to turn a small profit soon after reorganization, and by 1935 it was once again a healthy operation. The models of 1934—including the Commander Land Cruiser, with its 119-in. (3023-mm) wheelbase—were to thank for the turnaround.

The extent of Studebaker's revival can be measured by the engineering changes that took place, including the introduction of the eight-cylinder Commander line. The Commander Land Cruiser's engine displaced 122 cu. in. and, with a compression ratio of 6.3:1 within an aluminum cylinder head, produced 103 hp—the most powerful Commander to date. In addition to its capable engine, the Land Cruiser model—also available as part of the higher-output President series—featured new beaver-tail body styling inspired by the rakish Pierce Silver Arrow, the bodies for which had been built by Studebaker. The new body shell—an early peek into "fastback" modeling, highlighting the use of fleeting fender skirts and a Silver Arrow–like rear-window treatment—was an attempt to appeal to buyers in the moderate-price field. The new, more streamlined silhouette came with a recessed trunk, a more rounded grille, and taillights built into the rear fenders; elsewhere, Studebaker's trademark oval headlamps (in use since 1931) were supplanted by streamlined round headlamps with an innovative triple beam. The triple-beam headlamps could be adjusted via a control on the instrument panel that allowed the driver to select one of three beams best suited to the driving conditions and carrying load.

Standard for all Commander models were a Delco-Remy ignition, mechanical "steeldraulic" brakes, an improved-construction X-frame, and a Stromberg carburetor. The five-passenger Commander Land Cruiser was priced at $1220.

At $1220, Studebaker's Commander Land Cruiser was the top offering from the marque's mid-range lineup for 1934.

Aerodynamic body lines, rounded headlamps, and horizontal hood louvers distinguished the 1934 Commander Land Cruiser. Other features included deeply skirted fenders, V-shaped single-bar bumpers, and a streamlined trunk.

Left: Every angle and line on the Commander Land Cruiser emphasized the car's fluid body shape. Even the door handles were designed to convey movement.

Opposite: One of the highlights of the Studebakers of 1934 was this rather handsome clock-face speedometer.

OIL
GAS
0
10
20
30
40
80
90
100
180
212

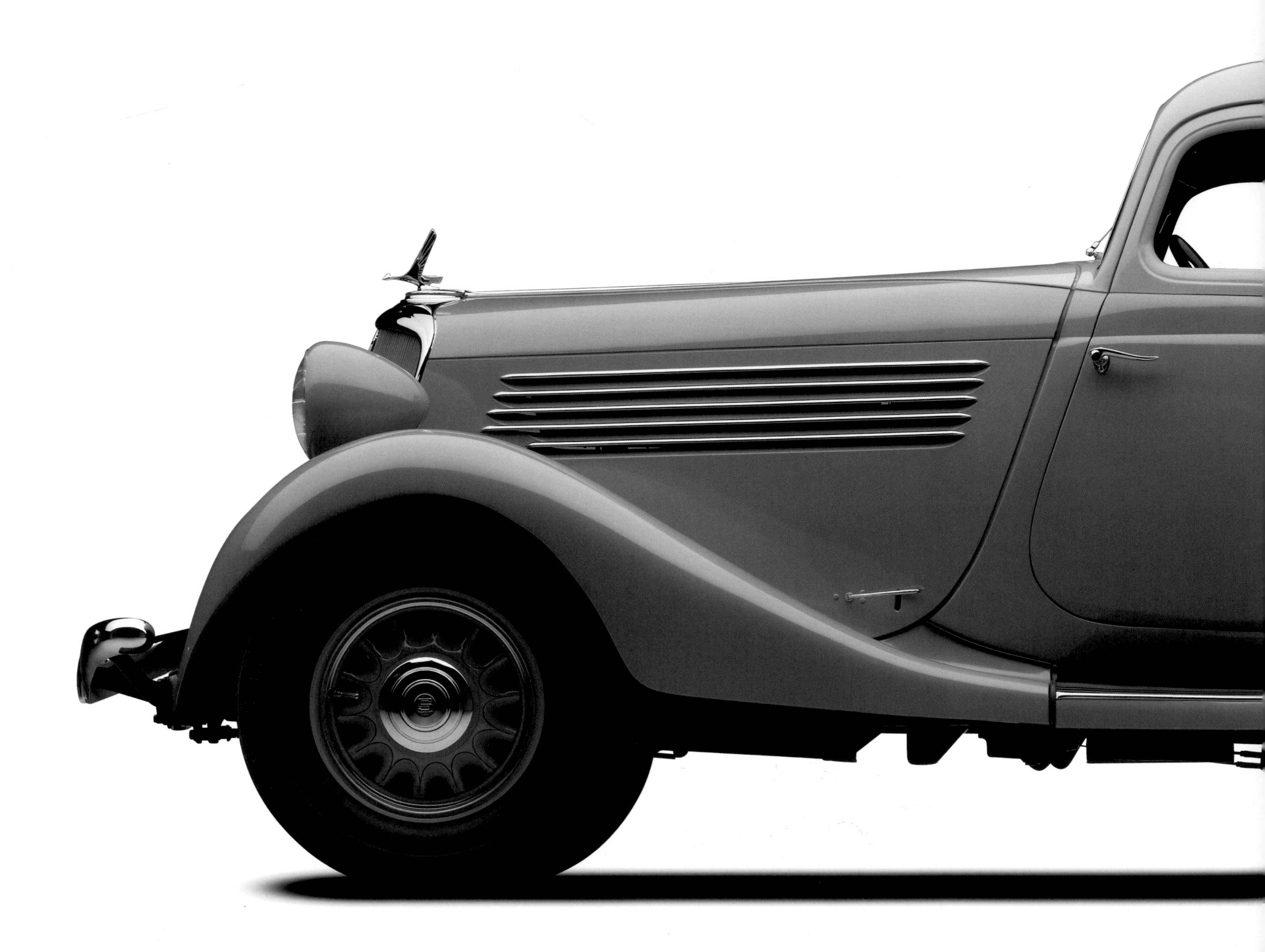

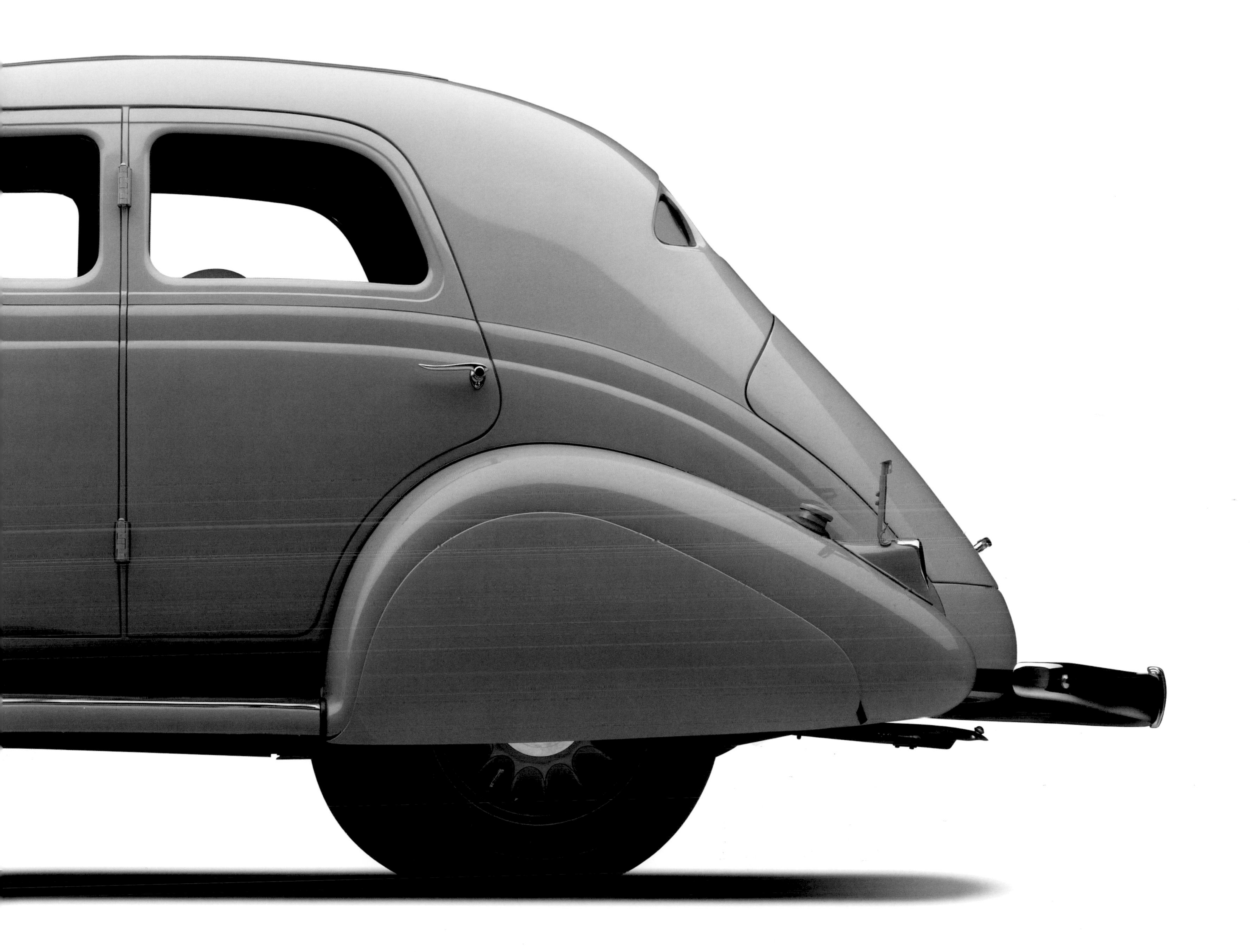

SUPER-CHARGED

1935 AUBURN 851 SPEEDSTER

It is easy to understand how one could be seduced by Auburn's speedster of 1935. The sleek body that tapered to a boattail did not belie the powerful eight-cylinder engine that sat under the hood. Assisting the seduction was racer Ab Jenkins, who completed a run averaging above 100 mph (161 km/h) at Bonneville Salt Flats in Utah in a completely stock SC (supercharged) Speedster, the first fully equipped American stock car to maintain such speeds over a twelve-hour period.

A genuine American auto legend, the gorgeous 1935 Auburn 851 Speedster was crafted by imaginative designer Gordon Buehrig.

Launched on January 1, 1935, the 851 sported a revised radiator grille and modified fenders that improved on the design of previous Auburns. Of the various body styles, the speedster, at the hands of master designer Gordon Buehrig, received universal acclaim, and is still considered one of the greatest designs of the American classic era. A basic 851 Speedster was offered alongside the SC version.

The speedster's features included a dual-ratio rear axle that incorporated a Warner Gear three-speed transmission, the result being six forward speeds. Drivers in the compact cockpit could pre-select gears by turning a dial under the horn button on the steering-wheel column. Each two-passenger car was virtually hand-built; at $2245, it was also a bargain. To remind passengers of its heritage—and as if to prompt them to hold on tight—each speedster came with a plaque on the instrument panel noting the speed at which it had been tested: just over 100 mph (161 km/h). The signature of either Jenkins or fellow test-driver Wade Morton also appeared on the plaque.

Sitting atop a 127-in. (3226-mm) wheelbase, the Lycoming straight eight, with a displacement of $279\frac{1}{4}$ cu. in., was capable of 150 hp at 4000 rpm, and could top 100 mph (161 km/h) with a Schwitzer-Cummins supercharger. In fact, that speed was "guaranteed" by the company for its 851 SC models. The Auburn 851 SCs were intended to lure customers into showrooms, but the effectiveness of this approach was limited. Out of a total of just more than 6300 Auburns of all types sold in 1935, approximately 500 were speedsters.

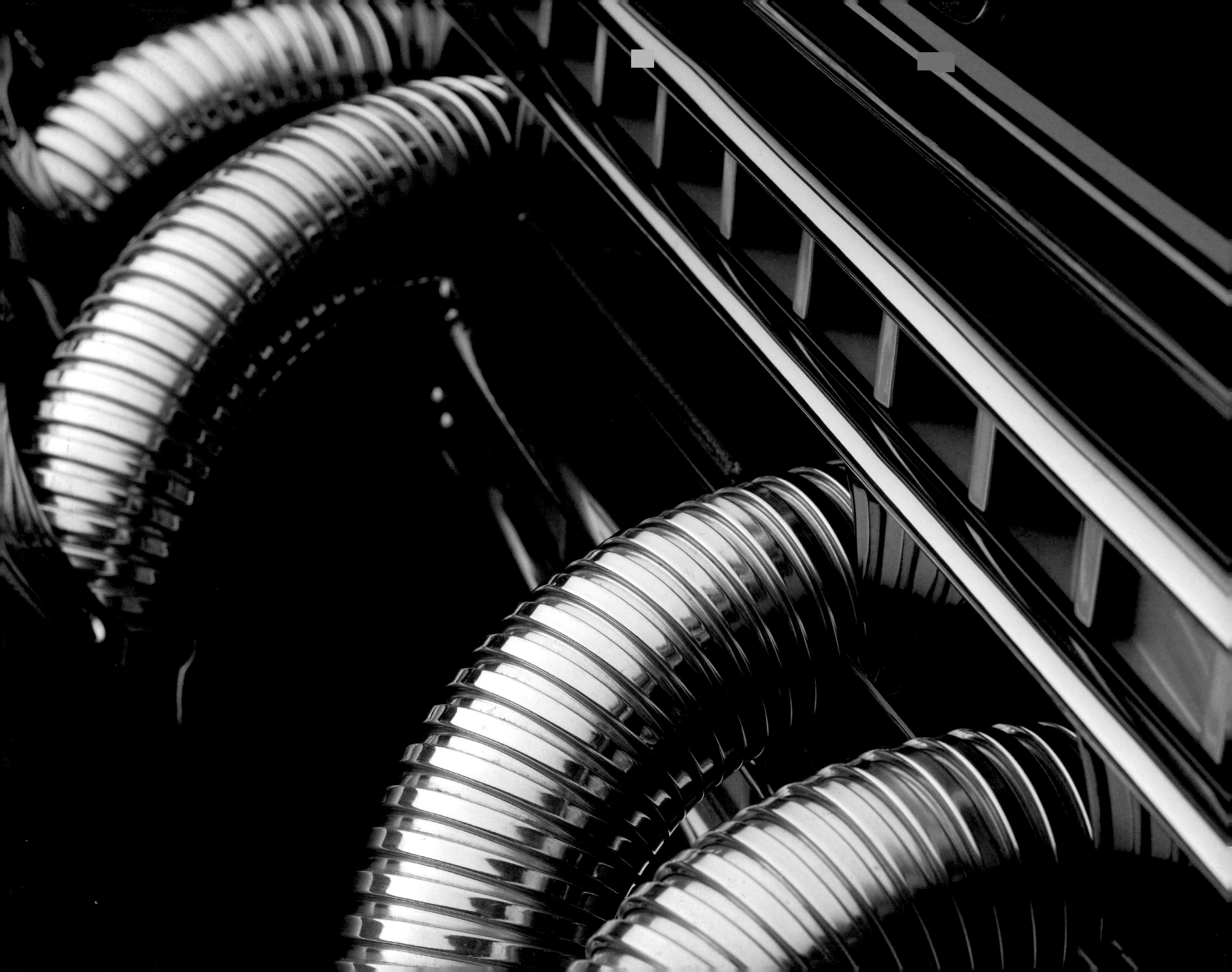

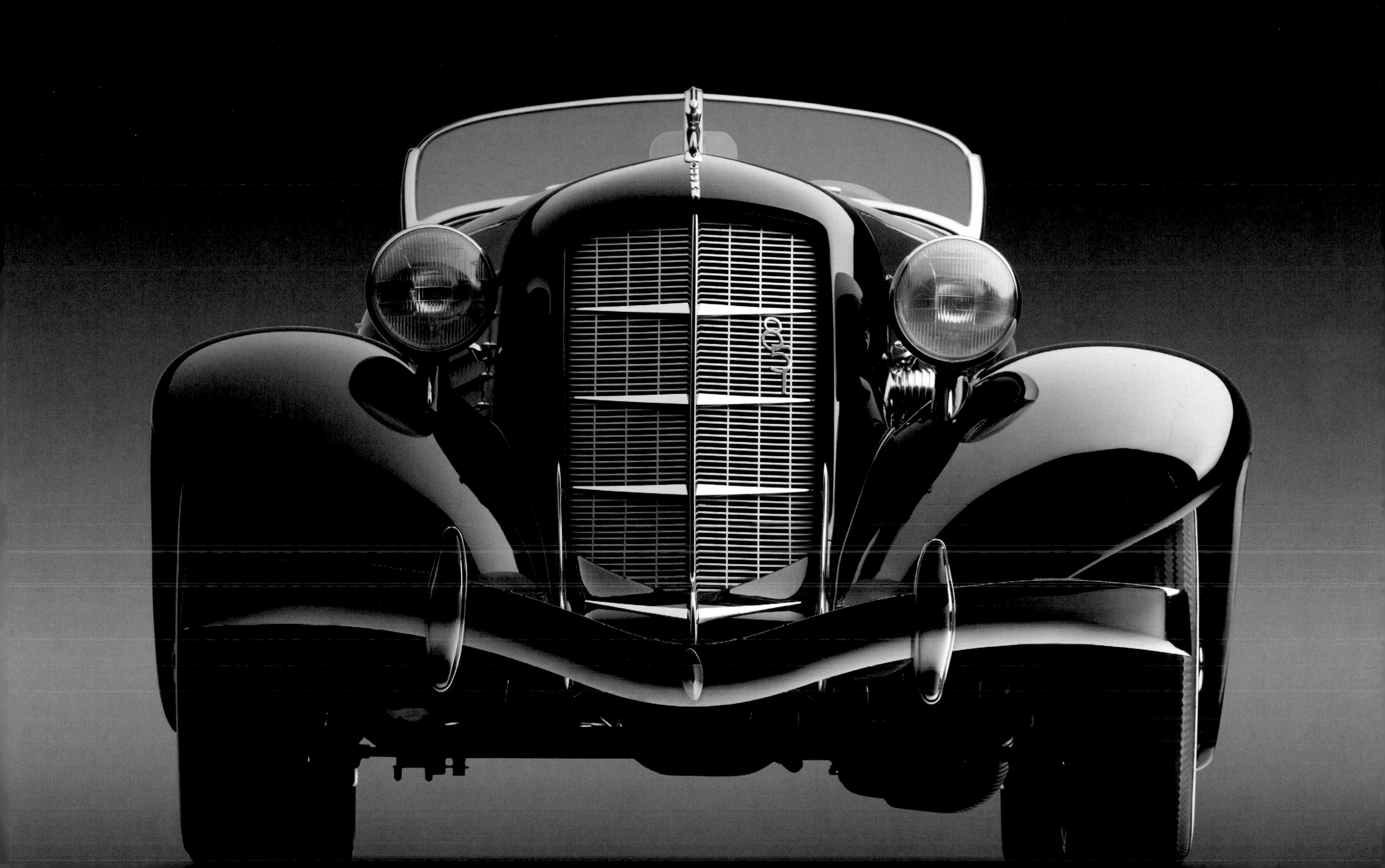

1935 NASH AMBASSADOR 8 SEDAN

Charlie Nash's interest in early automobiles brought him a long way, from an Illinois farm to the busy hubbub of Detroit. Nash was noticed in 1890 by William C. Durant, who would later form General Motors, and became a supervisor at Durant's Flint Road Cart Company. Nash co-founded Buick Motor Company, and in 1908 became Buick's president and general manager. Two years later, he was general manager of GM. Nash struck out on his own in 1916, however, and formed Nash Motors the following year. His expertise in management and knack for attracting the industry's most talented engineers quickly paid off.

Solid construction and handsome styling were Nash's strong points. The Ambassador body style was first used during the 1927 model year, when a specially trimmed club sedan form of the Nash Advanced Six was developed. At the time, it was the company's most expensive car. The Ambassador 8 came into its own model range in 1932 and was offered in various body styles.

Nash is best known for responding to public demand by building smaller and more economical cars. This trend was seen first in its downsizing of the Ambassador 8 series. The 1935 model year saw a complete re-styling—the introduction of Nash's "Aeroform" design—and a trimming of body styles. (A new two-door sedan was also added to the series.) The 1935 Ambassador 8 was built on a much shorter, 125-in. (3175-mm) wheelbase, using the previous and smaller Advanced Eight engine. For Nash, it was the end of large, classic cars seen in production from 1930 to 1934.

The new Aeroform design direction was touted in sales literature as giving Nash "Flying Power." The rear-end "slide" design, which sloped at a 35-degree angle, was especially noticeable on the Ambassador 8 Sedan. The price of this four-door, eight-cylinder car was $1165.

The year 1935 was the first in which Nash used all-steel bodies and hydraulic brakes in the construction of its cars. Sales reached 17,739 for all Nash models.

Nash autos largely lived up to the marque's slogan: "Give the customer more than he has paid for." The cars' interiors, for example, were handsomely done and strikingly modern.

The 1935 Ambassador 8 benefited from the introduction of Nash's fastback "Aeroform" styling. Of particular note were the sloping rear and streamlined fenders.

1939 LINCOLN ZEPHYR COUPE

In the mid-1930s, Lincoln and its dealerships were in need of a shot in the arm, and not just from an economic standpoint: the marque needed inspiration, a "halo" car. Such a car would have to be more than a traditional replacement for the Model K; it would have to be something revolutionary. That something was provided in the form of the Lincoln Zephyr of 1935–36. With its unit body design and rear engine (on the prototype), the Zephyr contributed to revamping the automotive picture of the immediate prewar period. From the beginning, it was a sophisticated vehicle for the upper classes.

Introduced in 1935–36, the singular Lincoln Zephyr was further refined in 1939. The various modifications included a larger grille with vertical bars.

The Lincoln Zephyr held a special place in the heart of Lincoln president Edsel Ford, son of Henry, who valued striking styling and smart design. The streamlined exterior—with sloping rear deck, curved side-window corners, and headlamps handsomely molded into the front fenders—was matched in terms of design by the car's accommodating interior motif. But backing the new model's flash-and-dash was state-of-the-art engineering. An L head V-12 engine was developed for the Zephyr from Ford's V-8, allowing the powerplant of the 1939 Zephyr to displace $267\frac{3}{8}$ cu. in. and produce 110 hp at 3900 rpm.

Continuing to stand out from the crowd thanks to a singular grille, the 1939 Zephyrs were even larger than before. Lower body panels also enclosed the car's running boards, and the bumpers, front and rear, were reshaped to give them a more rounded appearance. The cars also featured new hydraulic brakes. New on the inside in 1939 was yet another updated layout, with vertical pleating in the upholstery and the option to customize various components. Optional leather was offered in four colors.

The Zephyr succeeded in its role as the lower-priced luxury brand in the Lincoln lineup. Together with the Lincoln Mercury, the Zephyr bridged the gap between Ford's DeLuxe range and the high-end Lincoln K series. Calendar-year production for Lincoln Zephyrs in 1939 was 22,578 units.

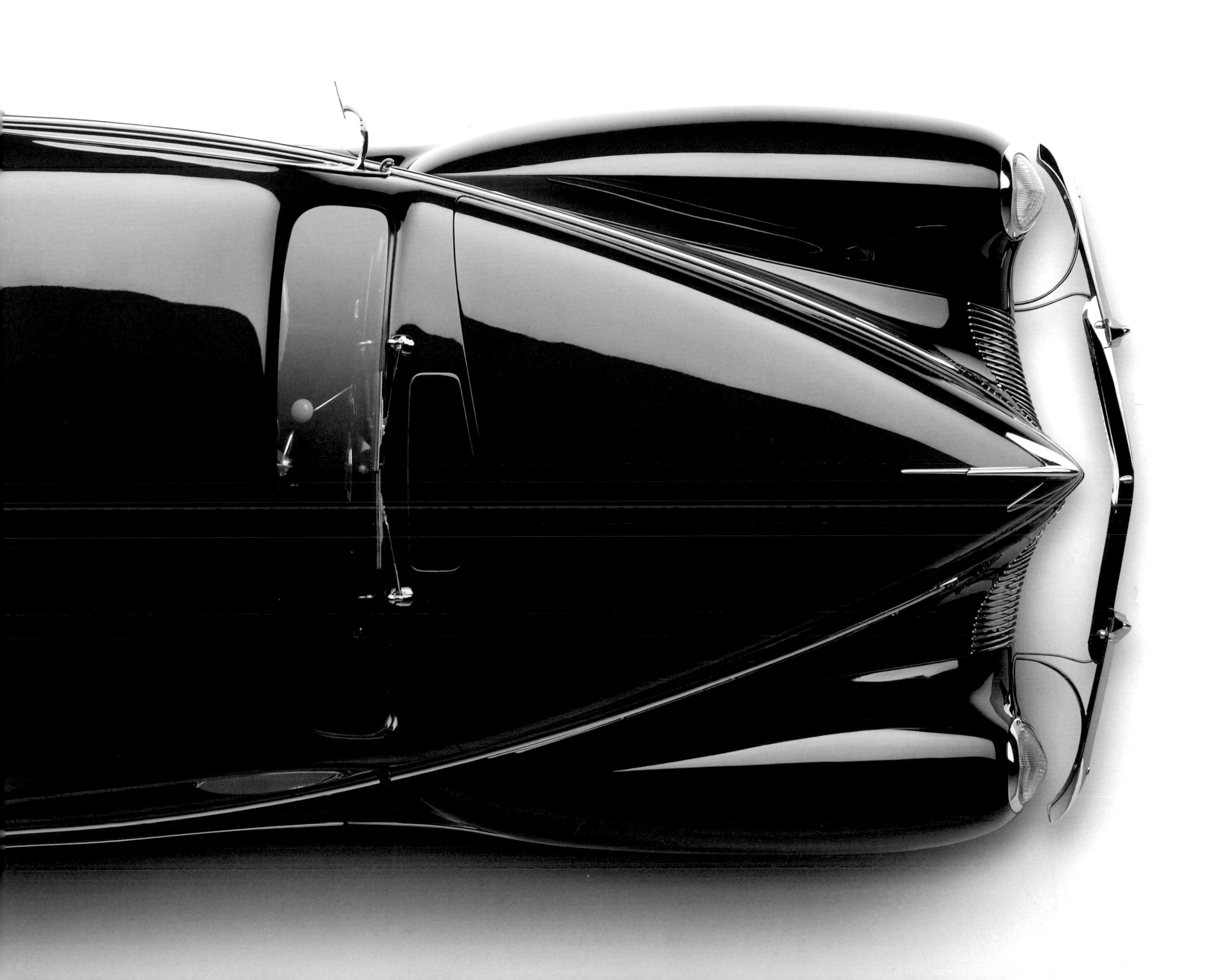

Named after the streamlined *Pioneer Zephyr* express train, the Lincoln Zephyr was powered by a V-12 engine. It was sold alongside the traditional Lincoln Model K until the K was dropped at the end of 1939.

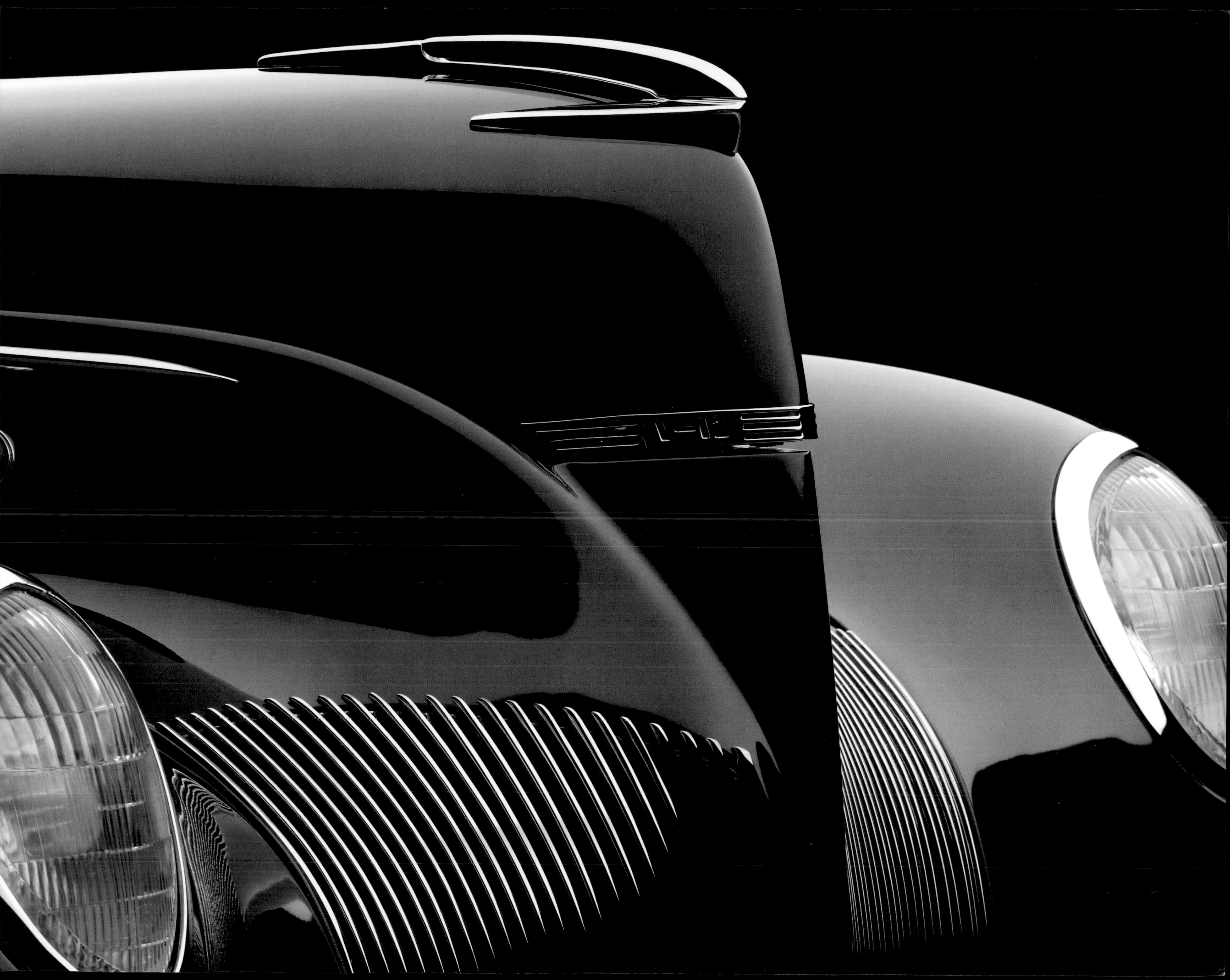

1940 LASALLE SERIES 52 CONVERTIBLE

On paper, LaSalle was Cadillac's failure. Despite resolute efforts to make the cars a success, production ceased in the summer of 1940. LaSalle's legacy, however, marks the dawn of modern car design in America. Beautiful lines, great performance, and a quality-car heritage made the first LaSalle of 1927 a real grab: the 303 of 1927–28 (see page 67) sold 26,807 units. During the first year of LaSalle production, Cadillac was obliged to build nearly 21,000 more cars than it had ever built before during any single model year.

The Series 52's narrow vertical radiator grille clearly marked it out as a LaSalle. All Series 52s came with Fisher-designed bodies on GM's new C-body platform.

Success would follow LaSalle. Another styling triumph equal to that achieved on the 1927 model was struck in the mid-1930s, with modern and dramatic body lines and smooth, clean, almost austere body planes replacing all traces of the old "classic" styling. The smoother look was relieved by such accents as porthole louvers on the hood sides and open, double-bar bumpers with telescoping spring mounts. Pontoon fenders and a streamlined body gave the car a much more substantial look, which had a heavy influence on all GM car lines during the 1930s. But the focal point of the new styling was a very narrow, vertical radiator grille. This slim-nosed look served as LaSalle's trademark through the model line of 1940. Bodied by Fleetwood, these cars were available in an "unlimited" choice of color schemes.

LaSalle went out with a flourish. The models of 1940, especially the Series 52, were the handsomest of the marque, with refined styling that featured new sealed-beam headlights nestled into the fenders, gently rounded "torpedo" forms, and restrained detailing. Prices for the nine-body line ranged from $1240 to $1895. The throat diameter of the dual-downdraft Carter carburetor was increased by 1/8 in. (125 mm), resulting in 130 hp for the final V-8. Refinements to the engine and chassis provided quieter operation and improved riding. Production in LaSalle's final year totaled 24,133 units, compared to 90,000 small Packards and 22,000 Zephyrs.

La Salle

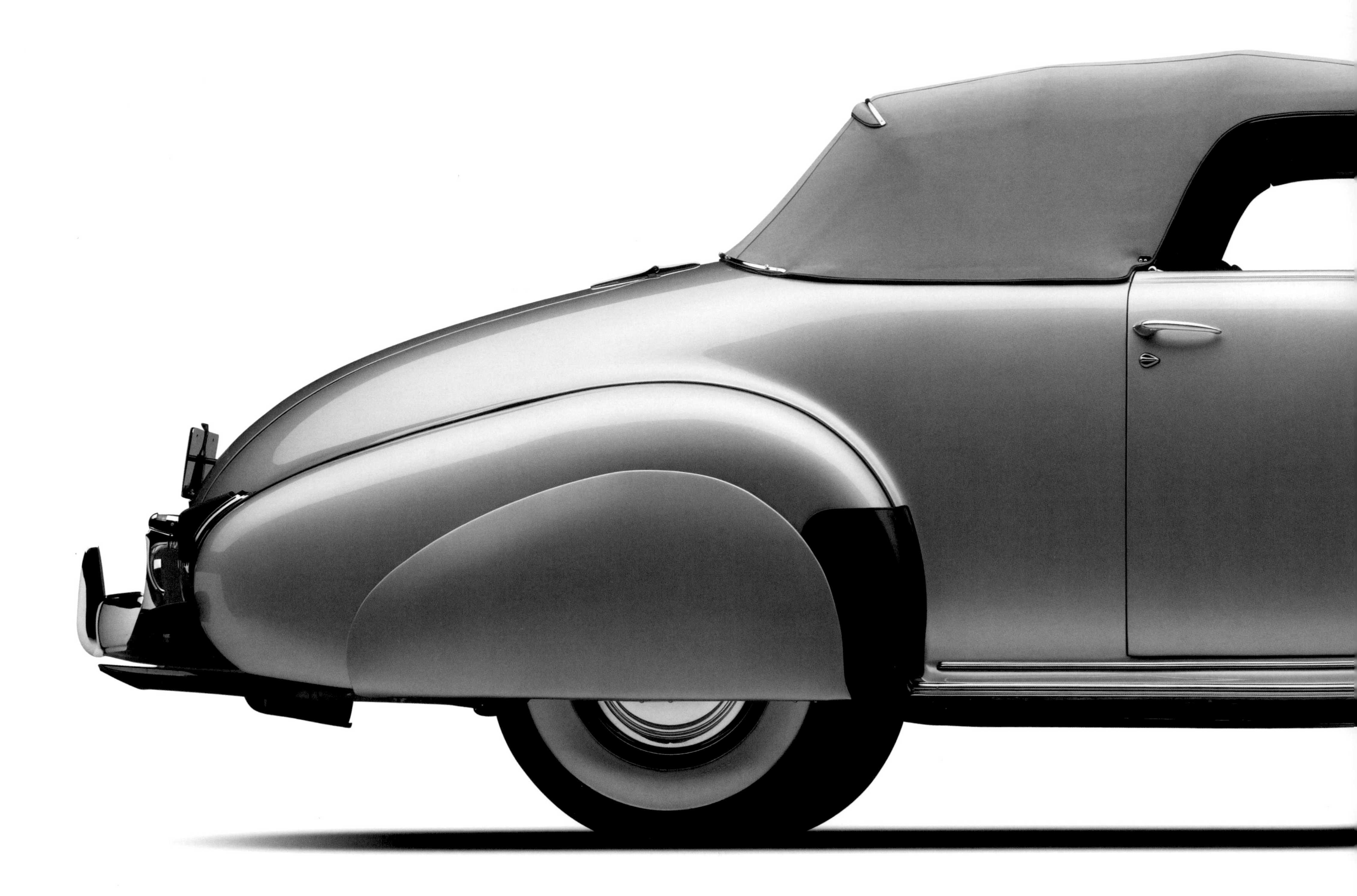

The Series 52 Convertible was known for its distinctive side louvers. Priced at $1535, only 425 were built for 1940, the final year of LaSalle production.

1940 LINCOLN CONTINENTAL CONVERTIBLE

The Lincoln Continental was conceived in the minds of Edsel Ford and Eugene "Bob" Gregorie. Edsel's request in 1939 for Gregorie to produce a one-off for his personal use, and to do so in strict confidence, resulted in a body design that would reverberate for years to come. Since 1913, Edsel had been attracted to European car design, and it was this interest that shaped Lincoln's offerings for high-brow socialites in America.

Priced at $2840, the Continental Convertible of 1940 was a production-worthy, top-of-the-line extension of Gregorie's custom Lincoln Zephyr for Edsel Ford. Mechanically, the Continental shared the Zephyr's 292-cu.-in. V-12 engine that developed 120 hp at 3600 rpm. But that's where the similarities ended.

Separating the Continental from its contemporaries was its ability to combine individuality with near-perfect proportions. The car was built on the standard 125-in. (3175-mm) Zephyr chassis, yet it was 7 in. (178 mm) longer and 3 in. (76 mm) lower. The Continental was easy to identify from a distance, with an overall length of 210 in. (5334 mm), a long hood, a comparatively short passenger compartment, and a distinctive Continental tire mount on the rear deck. Every model was largely handmade from modified Lincoln Zephyr components, with the "Lincoln Zephyr" legend seen on the horn button and hubcaps during the model's maiden year. In September 1940, the Continental name came into its own, and the Lincoln Zephyr moniker was dropped. By the end of the selling season that fall, 404 Continentals had been delivered to customers.

The first 1940 Continental, the Los Angeles show car, was presented by Edsel Ford to Mickey Rooney. In time, the Lincoln Continental was coveted far and wide. Architect Frank Lloyd Wright said the Continental was the most beautiful car ever designed, while author John Steinbeck noted that no other car "so satisfied my soul."

The venerable Lincoln Continental made its debut in Los Angeles. It was admired by many, including movie stars, influential architects, and renowned writers.

The Continental was based on a one-off vehicle custom-designed by chief stylist Eugene Gregorie for Edsel Ford's personal use. Favorable responses to the custom car guaranteed the development of a production version.

The Continental's long, elegant hood was designed to cover a Lincoln V-12 engine. The rear-mounted covered spare tire became the series' trademark.

Firestone
DE LUXE CHAMPION

1940 OLDSMOBILE SERIES 90 SEDAN

A noticeable sales slump hit the American auto industry in 1938, but Oldsmobile bounced back, moving from a 93,706-unit production low that year to an output of 158,560 in 1939. Another increase was in store for 1940, with production reaching 215,028 units, a figure that benefited from one of the country's most significant engineering feats to date: the Hydra-Matic transmission. Oldsmobile was the first to implement the new technology, adding it to its models of 1940 and calling it "the most important engineering advancement since the self-starter." That description wasn't far from the truth. In advertisements for its lineup of 1940, the General Motors division boasted, "No Gears To Shift . . . No Clutch To Press!"

With its lowered hood and grille, the Series 90 was Oldsmobile's premium offering for 1940.

Earlier, in 1937, Oldsmobile had provided the Automatic Safety Transmission, a semi-automatic unit requiring a conventional friction clutch for stops and starts. The Hydra-Matic advanced the technology to the status of fully automatic, with a four-speed transmission combined with a highly efficient fluid coupling—"unquestionably the beginning of a new trend in automotive development," noted *MoToR* magazine.

One of the few models of late 1939 to carry the new transmission was the Series 80, a 257-cu.-in., 110-hp eight-cylinder. The following year, the "80" was dropped and the "90" was offered, with the same engine as before but this time on a 124-in. (3150-mm) wheelbase. The Series 90 was Oldsmobile's top tier, with easy-on-the-eye styling and power to boot, all for a little more than $1000.

Exclusive treatments found on the Series 90 included front-fender chrome trim, optional two-tone paint schemes, and a deluxe steering wheel. A new offering on the 90 was the use of foam padding over the seat springs, making for an even smoother ride. All Series 90s of 1940 featured walnut grain-finished instrument panels.

The touring sedan was the top-selling body style of the series. Fittingly, the millionth Oldsmobile was a 1940 Series 90 Sedan.

The Series 90's distinguishing features included a semi-notchback profile, blanked-out rear quarters, and vent windows in the rear doors.

1941 CADILLAC SIXTY SPECIAL

For Cadillac in 1938, two events in particular were newsworthy affairs: the development of a new V-16 engine, and the arrival of the outstanding Sixty Special. The Sixty Special sat atop a 127-in. (3226-mm) chassis and featured a new frame and distinctive styling. What made the car special was its all-new, notch-backed body with convertible-style window styling; it was also smaller than its recent V-8 predecessors. An alligator hood—the first such hood to be used on a production car, together with the one found on the LaSalle of that year—was also part of the Sixty Special design.

As the LaSalle had been Harley Earl's baby, so the Sixty Special was Bill Mitchell's. Mitchell was the chief designer in the Cadillac studio, and would go on to replace Earl as head of GM Styling. Mitchell's creation started out as a LaSalle but ended up as the Sixty Special, the first sedan without a running board. The car was the fulfillment of a desire to simulate a four-door convertible in a sedan body. The base price for the first full-year production model was $2090, almost $1000 less than the larger Model 75.

A total of 4000 Sixty Specials were produced for 1941. That year's model benefited from an across-the-line horsepower increase, to 150 hp at 3400 rpm, and a compression ratio of 7.25:1 for its 346-cu.-in. V-8. The frame on all series models was made 40 percent stiffer, and the offerings for 1941 included the Special as a five-passenger Fleetwood sedan available with a "sunshine roof" or glass chauffeur division. The models of that year had a maximum speed of 100 mph (161 km/h), and could accelerate from 0 to 60 mph (97 km/h) in 14 seconds.

The Sixty Specials of 1941 were the first to be made available with the optional four-speed Hydra-Matic automatic transmission. Other available options—most of which were uncommon at the time—included a radio, fender skirts, driving lights, mirrors, windshield washer, back-up lights, integrated headlamps, and air conditioning, known then as "Weather Conditioning."

The brainchild of chief designer Bill Mitchell, Cadillac's Sixty Special set a new design standard and became the most influential of the company's prewar models.

In 1941, the Sixty Special was the flagship of the Cadillac fleet. That year also saw the first appearance of the company's famous "eggcrate" grille.

The Sixty Special was the ultimate in luxury. Its crisply formal looks were enhanced by fenders with squared-off trailing edges that harmonized nicely with the car's contours.

1947 CHRYSLER TOWN AND COUNTRY

No sooner had the Second World War ended than the race in America to produce automobiles was on once again. Some manufacturers were quicker than others, while some reached back to what was on the drawing boards for the aborted 1942 model year. Chrysler drew on its Town and Country to lead it into the immediate postwar period. Originally, the Town and Country had evolved from a model conceived in 1941 by Chrysler stylist Buzz Grisinger, who had combined the roof of a Chrysler seven-passenger sedan with the windshield, cowl, and fenders from a Chrysler Windsor. The resultant station wagon was decked out with contrasting light-and-dark woodwork.

Between 1946 and 1948, Chrysler sold more non-wagon woodies than any other manufacturer. The most famous was the 1947 Town and Country convertible.

The postwar Town and Country was a revolutionary four-door "personal car," an extension of Chrysler's station-wagon focus. The lineup consisted of models that featured either six- or eight-cylinder engines. Most Town and Country models used the six-cylinder Windsor platform, whereas all convertibles were on the longer, New Yorker chassis, with its 135-hp, 323-cu.-in. straight eight. The Town and Country's most distinguishing characteristic was its all steel body, the lower portions of which were covered in white-ash and mahogany panels. The rounded body lines were balanced with a dazzling "harmonica" grille.

On the inside, handsome styling added to the Town and Country's distinctive place among Chrysler's offerings. Convertibles came with all-leather components, pleated-leather inserts and bolsters, and Bedford cord upholstery.

In 1947, dual spotlights and rearview mirrors came as standard, as did a roof-top luggage rack and roof bars. The plentiful exterior woodwork, along with a smooth ride and sporty elegance—inside and out—made the 1947 Town and Country models tops in terms of non-wagon "woodies" sold before 1950.

Chrysler began producing its distinctive lineup of wood-trimmed luxury vehicles bearing the Town and Country name in 1946. The 1947 convertible was one of the most coveted.

Left: The Town and Country's interior featured all-leather components, wood paneling, and Bedford cord upholstery.

Opposite: Structural ash framed the car's bodysides and rear deck. Inserts were genuine mahogany through mid-1947, then realistic Di-Noc decals.

Cadillac

1948 CADILLAC SERIES 62 CLUB COUPE

For 1948, Cadillac's Series 62 models—which replaced the mid-sized Series 70 in 1941—were restyled with an all-new, full-width approach that featured straight-through slab sides and new, drum-type instrument clusters. The coupe (also known later as the "sedanette") was especially handsome, with integrated rear bumpers and a gasoline-tank fill-cap hidden beneath a specially hinged taillight. Part of the first postwar redesign for GM's luxury division, the restyled Series 62s were among the most graceful cars of the era. Most models were equipped with the game-changing Hydra-Matic transmission, and low-pressure tires were now standard equipment.

More influential, however, was the presence of tail fins, an idea that GM styling chief Harley Earl carried over from his fascination with the Lockheed P-38 Lightning fighter aircraft from the Second World War. Of all the Lightning-inspired ideas seen on sketches—pontoon fenders, propeller-shaped noses, and aircraft-like cockpits among them—tail fins was the one to make it on to the models of 1948. While some Cadillac dealers initially balked at the presence of what many regarded as quirky "nubs" at the rear of the car, it was soon evident to all that Earl was on to something. The tail-fin rage had begun.

The Cadillacs of 1948 were the last to use the firm's rugged L-head V-8. The 346-cu.-in. engine was more than capable of powering the two-door Series 62 Club Coupe, which, compared to the previous year, had a shorter wheelbase—now 126 in. (3200 mm) rather than 129 in. (3277 mm)—and a lighter body weight. The flathead V-8 produced 150 hp.

In 1948, a total of 4764 Series 62s were produced, with a basic price of just under $3000. The new styling caught on, and the car's unit sales for 1949 increased by nearly 3000.

The Series 62 was the first Cadillac to reenter production following the end of the Second World War. It is also credited as the first car to feature tail fins.

Dynaflow
Roadmaster

1949 BUICK ROADMASTER

The 1949 Series 70 Roadmaster lineup is one of Buick's most enduring and beautiful range of cars. As Buick's largest and most prestigious automobiles, Roadmasters were powered by a 320-cu.-in. version of the company's famous overhead-valve, straight-eight engine that produced 150 hp. The Buick Dynaflow automatic transmission, which had been introduced the year before, came as standard, as did front-fender portholes known as "Ventiports." These style elements, popularly attributed to designer Ned Nickles, heralded the introduction of an aircraft-design theme that was seen on other General Motors products.

The 1949 Roadmaster was built on Buick's longest wheelbase. It was the carmaker's premium and best-appointed model.

For 1949, Buick produced nearly 400,000 cars, a record thanks in part to the success of the Roadmaster. Postwar Buicks were based on GM's new C-body platform, enabling the division to produce the widest selection of models in its history, highlighted by the rakish and luxurious Roadmaster, which sat on a long, 126-in. (3150-mm) wheelbase. Together with Buick's Super series, the Roadmaster received body-length fenders, a panoramic curved windshield and rear window, and one-piece wraparound bumpers. The Roadmaster Riviera Hardtop Coupe introduced "hardtop convertible" styling, which essentially removed the B-pillar—a rarity in concept and design. Only Chrysler had done something similar, with its Town and Country cars in 1946, yet only seven prototypes had been produced. Buick made more than 4000 hardtop Rivieras in 1949.

Staying true to its name, the Roadmaster—a moniker that first appeared on Buick automobiles in 1936 to note engineering improvements and design advancements—remained the firm's highest-end offering for years to come. Many design elements endured over the course of time, including a bright-metal side decoration called the "sweepspear," which appeared on some models. The sweepspear began in the front fender as a slim horizontal molding, and widened as it swept down in a curve along the doors, eventually ending in an upsweep above the rear-wheel openings. As with other elements of the early Roadmaster, this feature, in time, changed in detail but not in concept.

HEATER
DEFROSTER

The devil is in the detail . . . From the dashboard controls and the speedometer to the aircraft-inspired taillights, nothing escaped the careful attention of Buick's designers.

FUTURAMIC

1950 OLDSMOBILE HOLIDAY DELUXE COUPE

The first postwar restyle at Oldsmobile, along with a similar restyle at General Motors' Cadillac division, was scheduled for late in the 1948 model year, with universal application to all GM models by 1949. The task was entrusted to Harley Earl and his talented crew of stylists. What they came up with—lithe, low, and lovely compared to the bulbous shapes of the past—would last GM through 1953, and in some cases beyond. Oldsmobile's new design direction was given the proper name of "Futuramic."

Stylish, modern design distinguished the 1950 Olds Holiday coupe, one of the first Oldsmobiles to showcase the company's new "Futuramic" design direction.

The new look was introduced on Oldsmobile 98s of late 1948, in both convertible and sedan form. Apparently, GM felt that its upper lines would be best served by the introduction of the new style pre-1949; as a result, Oldsmobile beat several rivals, notably Mercury, by as much as a model year. At the same time, Olds could offer an immediate counter to the "step-down" design of 1948 created by the burgeoning Hudson Motor Car Company. While not as advanced as Hudson's step-down models, and conventional in both engine and chassis, the Futuramic 98—soon available in Holiday Deluxe Coupe form—looked properly up-to-date and far more "modern" than what had come before.

With the models of 1949 and the marque-wide application of Futuramic styling came an Oldsmobile development that would later be considered a milestone: the introduction of the Rocket V-8 engine, designed by Gilbert Burrell. This powerplant shares with a similar but separate Cadillac design the honor of being the first modern, high-compression, overhead-valve V-8. The new engine, developing 135 hp at 3600 rpm in its original state, was immensely strong. Originally, the Rocket was intended for the heavy 98, but division general manager S. E. Skinner, in a flash of enlightenment, saw to its application in the lighter, short-wheelbase 88 as well. This gave rise to a NASCAR racing legend, the powerful Rocket propelling the nimble 88 to several stock-car championships.

HUDSON

1951 HUDSON HORNET SEDAN

Partly an answer to Oldsmobile's near-dominance of stock-car racing, and partly a response to the general scramble for higher compression ratios in lighter car bodies, the Hornet was also Hudson's attempt to steal the limelight. The Hudson Hornet debuted in 1951, with a new, 308-cu.-in. L-head six-cylinder engine that developed 145 hp at 3800 rpm and 257 lb ft of torque—the latter figure a 30-percent improvement over Hudson's eight-cylinder engine. The new powerplant gave a 10-percent increase in top speed, reduced the standing quarter-mile time by about a second, and considerably improved passing acceleration. Ample in its displacement, operating well within its capacity, and with a chrome alloy block, it was very tough and durable. It was also the last big six ever to be built in America.

Drowning out some drivers' displeasure with the engine's drop in refinement—notably, a less smooth and noisier operation—was the announcement by Hudson that it would go racing with its new Hornet. The program was directed by Hudson engineer Tom Rhoades, who led the charge in 1951 against six other manufacturers, all of which were competing with models more powerful than the Hornet. Yet, by the end of the NASCAR racing season that year, Hudson stood at third in the list with twelve wins, behind Oldsmobile and Plymouth. In 1952, although the Hornet was eighth in horsepower ratings compared with its rivals, it ran first in twenty-seven out of thirty-four NASCAR Grand Nationals, leaving Oldsmobile and Plymouth to tie for second place with three wins apiece.

Hudson gained enormous free publicity with these showings, and the loyalty of Hudson followers was intensified. In 1952, an editor at *Speed Age* queried the value of Hudson's racing program. Over the following months, the magazine received an avalanche of 3000 letters in response, three-to-one in Hudson's favor.

The 1951–53 Hornet is the most remembered Hudson of the postwar years, an all-time industry great.

Left: Surprising handling and quality engineering combined with luxury to make the Hornet a one of a kind. The body badge on the front quarter panels featured a rocket dissecting a capital "H"; its trail ran the length of the car.

Opposite: With the Hornet, plenty of car could be purchased for one's money; prices ranged from $2543 to $3342. On the track, it was virtually unbeatable through 1954.

HUDSON

1953 CADILLAC ELDORADO

The immediate postwar period in America was filled with car buyers anxious for something new in terms of style, design, and performance; following the end of hostilities, the public soon tired of staid, warmed-over models. Enter Cadillac, the General Motors division that had already established itself as a producer of luxury automobiles and a trendsetter in design. As Packard was known for conservative styling and engineering in the fine-car field, so Cadillac was heralded for its flashy approach. Never was this truer than with the introduction of the Eldorado in 1953.

GM design chief Harley Earl had championed the wraparound windshield on earlier concept models, and it was with this design feature in mind that he approached Cadillac general manager Don E. Ahrens in a bid to build an unprecedented luxury convertible. The idea was for the Eldorado to enhance the marque's prestige, with the special windshield setting it apart from the Cadillac Series 62 Convertible, on which it was based. The panoramic windshield was fitted with unique movable wind wings. Other differentiation ensued, including a chassis that sat 1 in. (25 mm) lower than standard, a dropped beltline that was notched above the vertical rear-fender scoop, and a special Orlon fabric top in either black or white that disappeared into a recess covered by a body-color metal panel.

The Eldorado's interior was set off by many elements. A modified instrument panel with a leather-finished crash pad sweeping back from the windshield and on to the door panels combined with chrome aplenty. The steering wheel came with special simulated-leather plastic handgrips, and door and interior quarter panels were garnished with wide chrome pieces and chrome kick panels at the sill. Seats were all-leather with special ribbed inserts.

The extras found on the Eldorado justified the price: $7750, more than $3600 higher than the standard Series 62 Convertible, and more than $2000 higher than the Fleetwood 75 Limousine. The Eldorado was now America's most expensive dream car.

The Cadillac Eldorado debuted in 1953 at General Motors' Motorama auto show. Billed as a dream machine, it certainly lived up to expectations.

The Eldorado's low, sleek body was distinguished by a notched beltline and a flush-fitting metal cover to conceal the folded cloth top.

One of the Eldorado's most distinctive features was its new "panoramic" wraparound windshield, developed by GM design chief Harley Earl. It would prove to be an extremely influential design.

1953 CORVETTE

By 1951, General Motors design chief Harley Earl was thinking seriously about a low-priced sports car. A year later, he asked staff designer Robert McLean to draw up plans for such an automobile. McLean saw it as a long, low roadster akin to the British sports cars that were captivating young American males and females alike. From McLean's sketches, Earl understood that special engineering and production were needed, and his go-ahead was given. The project was kept secret from the majority of GM staff.

Chevrolet engineers adjusted standard components to give the new car the performance it required, while the stylists explored the potential of fiberglass as a body material. In an astonishingly fast twelve-month turnaround, from concept to production, a small pilot assembly line in Flint, Michigan, churned out the first 300 Corvettes between June and December 1953. (Production eventually moved to a permanent home in St. Louis, Missouri.) In those early days, the bodies were all hand-built. It was truly a time of experimentation, with Chevrolet engineers and designers learning on the job as America's first mass-produced sports car came to life. All models of 1953 were genetically identical, with Blue Flame six-cylinder engines, Powerglide transmissions, white paint, and red upholstery.

Satisfying demand was an early problem. With so many people eager to purchase the first Corvettes, Chevrolet decided to sell them to high-level associates of GM, civic leaders, and other public figures. Despite the model's popularity, sports-car enthusiasts cried foul, as the purist's definition of a sports car held that it must have a manual transmission. Available with only an indulgent automatic, the Powerglide, combined with an underpowered six-cylinder (underpowered, that is, relative to European offerings of the time), the Corvette did not meet this definition. That would change within a couple of years, with the introduction of a manual transmission and V-8 combo.

The Corvette marked a turning point in American road performance and styling. The marque remains one of the country's most endearing.

An automotive milestone, the Corvette came from the desire to build, finally, an American sports car to compete with the Europeans.

50 60 70 80 90
120 130 140
BRAKE
PULL DEF
CHEVROLET

Opposite: The first 300 Corvettes were all hand-built, with Polo White exteriors and Sportsman Red interiors.

Right: The innovative use of fiberglass as a body material meant that the Corvette's designers could experiment with more complex shapes.

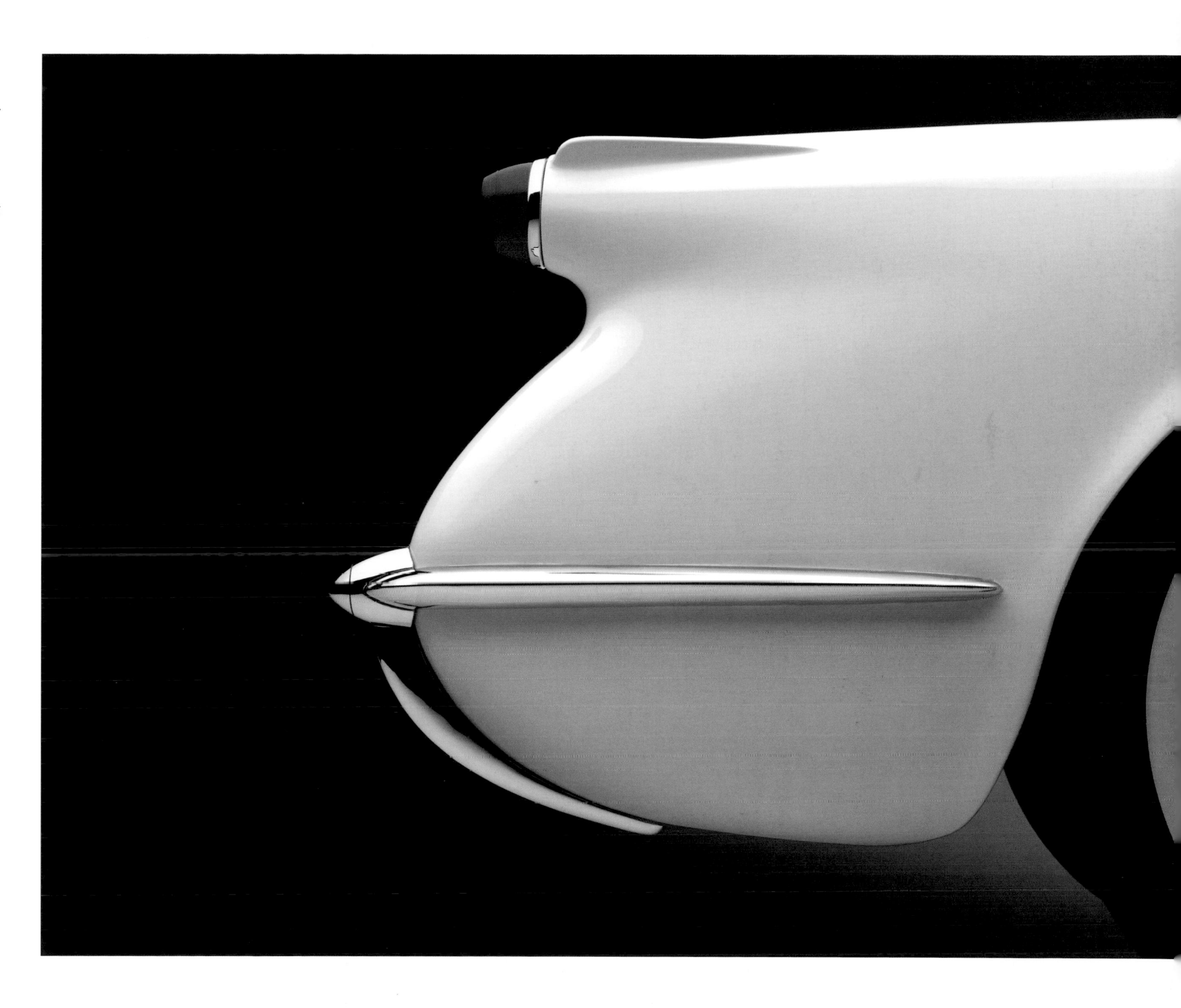

The first Corvette rolled off the assembly line just six months after the vehicle's public unveiling at GM's Motorama of January 1953. The influence of British sports cars was clear.

Notable among the Corvette's many distinguishing features were rocket-like rear fenders with small fins and a shadow-box license-plate housing.

1953 STUDEBAKER COMMANDER STARLINER COUPE

In 1951, Studebaker sales dropped to 205,000 units, as the company desperately held on to 4 percent of the market. Despite introducing a new overhead-valve V-8—a fine engine—the following year, sales dropped again, resulting in a market-share loss of 0.2 percent. It was obvious that the public had tired of a style that, although once radical, was now merely a facelift. Studebaker management reacted by considering a new design prepared by Bob Bourke, then chief designer for the Raymond Loewy group. The coupe design would come to be known as the "Loewy coupe."

The 1953 "Loewy coupe" represented the company's first entirely new body style since 1947. The pillarless Starliner edition was particularly noteworthy.

Bourke's design was longer, wider, and—at 55 in. (1397 mm) tall in steel-roof form—lower than the previous Studebaker models, while the overall style was a gorgeous blend of the best features of American and European design. After some arguments and inter-company concessions, Studebaker accepted Bourke's coupe and a sedan derivation. The production coupe was 2 3/8 in. (60 mm) taller than the prototype, but only 2 in. (51 mm) taller than the contemporary MG TD British sports car with its roof up. The sedans, at 60 1/2 in. (1537 mm) tall, were not low enough for Loewy, but management insisted on increased headroom. Even so, they were still 2 in. (60 mm) lower than previous models, and remained handsome designs.

Studebaker sedans used a 116 1/2-in. (2959-mm) wheelbase, while the coupe hardtop used a longer, 120 1/2-in. (3061-mm) version. Design proportions heavily favored the longer-wheelbase models, and the public went for them with considerable enthusiasm.

The V-8 was used in the medium-priced Commander versions, which, in the case of the pillarless Commander Starliners of 1953 and 1954, best represented the era's notion of combining simplicity, refinement, and beauty. When the "Loewy coupe" first hit the streets, a sloping front hood, wheel openings that completely exposed the tires, and minimal chrome trim created a clean overall design. Advertisements proclaimed it "The new American car with the European look."

The 1953 body shell was one of Studebaker's most widely used designs. The Hawk carried it until 1964.

SKYLARK

1954 BUICK SKYLARK

The Buick Skylark rolled out in 1953 as part of the General Motors division's golden anniversary. The limited-production Skylark, a $5000 sports convertible, was a distinctively styled model with circular wheel cutouts, imported wire wheels, a delicate rendering of the "sweepspear" side decoration introduced on Buicks in 1949, a dropped beltline, and an overall height 2 in. (51 mm) lower than other production Buicks. It was both a harbinger of the personal luxury automobile to come and a hearkening back to the prewar custom car. It was also the first time that GM had taken one production model—in this case the Roadmaster convertible—and reworked it into another. That first year, the Skylark enjoyed a successful production run of 1690 units.

The Skylark heralded the arrival of the personal luxury automobile. Buick offered the car with almost all available optional equipment fitted as standard.

The Buick Skylark was restyled for 1954, with elongated wheel cutouts and the option of having the interior in a different color to that of the body. All production Skylarks were built as two-door convertibles and had the same luxury equipment as the models of 1953. The trunk of the new Skylark was sloped into a semi-barrel shape, and taillights were housed in chromed fins that projected from the tops of the rear fenders. Now based on the shorter, all-new Century/Special chassis instead of the Roadmaster/Super chassis, the Skylark continued to share the Roadmaster and Century powertrain, which had the highest output in the 1954 Buick lineup. The high-compression Fireball V-8 engine displaced 322 cu. in. and produced 220 hp.

Although it eventually led to the Skylark's rarity, the restyle, as well as the high sticker price, adversely affected sales. Compared to the previous year's Roadmaster/Super based model, the public perceived the Century/Special-related Skylark of 1954 to be a step down in terms of rank. Poor sales led to the car's demise at the end of the 1954 model year, and it would be 1961 before the Skylark name reappeared on Buicks.

BUICK

Opposite: With its wide, baleen-like radiator grille, the Skylark was built to celebrate Buick's fiftieth anniversary. It was offered only as a convertible.

Right: Striking features could be found inside and out, including a tapered rear deck with standout chrome fins.

Thunderbird

1955 FORD THUNDERBIRD

Ford intended its Thunderbird to be a Corvette killer. And it almost succeeded. Capitalizing on America's growing lust for sports cars, Ford seized on what many people regarded as the weak points of Chevrolet's Corvette of 1953–54 (see page 170): its lack of power and poor construction. Introduced in 1955, the Thunderbird—a two-seater vehicle with a detachable fiberglass roof—was powered by a 292-cu.-in. V-8 engine that produced almost 200 hp. Both manual and automatic transmissions were available.

Besides the get-up-and-go requisite for a sports car, the Thunderbird offered extras that were missing from the Corvette, including roll-up windows. In fact, Ford used such "missing" extras for its marketing direction, propagating the car as a luxury model for personal use rather than a sports car.

The personal-luxury tact was more than hype. The T-bird's steering was smooth, allowing for maneuverability on busy roads, while the suspension included a ball joint at the front, offering a plush ride. The interior also gave the vehicle an elegant appeal, with comfortable seats and adequate legroom. A generous trunk provided plenty of luggage space for long trips.

The Thunderbird was also available as a soft top, priced at $2765; the basic price for the hardtop was $2695. Another popular option was a removable hardtop with circular portholes.

With such style, comfort, convenience, and power, the Thunderbird outsold the Corvette in the 1955 model year by a margin of 23 to 1. Franklin Q. Hershey, formerly with General Motors, had been hired by Ford in the early part of 1952. It was he—together with Lewis D. Crusoe and George Walker—who had developed what would become an American icon.

The Thunderbird of 1955 was Ford's response to Chevrolet's Corvette. A two-seater with a removable fiberglass top, the Thunderbird was a resounding success.

FORD

Opposite: Following a successful maiden year, few changes were made to the car in 1956. The most significant was the use of a Continental-style spare tire to increase storage space in the trunk.

Below: The sleek and athletic Thunderbird was a winning combination of European style, American comfort, and the power of a sports car.

1956 LINCOLN CONTINENTAL MARK II

In July 1952, William Clay Ford, son of Edsel, was named manager of the newly established Special Product Operations, a small coterie of engineers and stylists responsible for some of Ford's most intriguing products. One such product was the Lincoln Continental Mark II, a project so secret that it was not until after the car had been introduced that the public was even aware of the men responsible for its design: Gordon Buehrig, John Reinhart, Bob Thomas, and Harley Copp. Ideas for the car were gathered both from these top independent designers and from the Lincoln styling department.

The Mark II made its public debut at the Paris Auto Show of 1955, seven years after production of the first Continental had ceased. It marked the return of one of the great names of the American automotive world.

In October 1954, William Ford formally announced the impending launch of the Mark II. One year later, the model made its public debut at the Paris Auto Show. It was the much-anticipated revival of a grand name: production of the Continental, an evolution of the Zephyr (see page 125), had ceased in 1948.

Parental heritage was an expected attribute of the new Continental, and that is exactly what enthusiasts got. The Continental tire, the short rear deck, the long and low hood—all were part of the package. The perfect proportions that characterized the first Continental (see page 135) now marked the new version, except with a decidedly modern elegance. Unlike the first iteration, the Mark II shared neither frame nor components with any other Ford or Lincoln. Overall, it measured 216 in. (5486 mm) long, 78 in. (1981 mm) wide, and 56 in. (1422 mm) tall, a height that was both aesthetically appealing and functional. The lower center of gravity, coupled with a specially designed 126-in. (3200-mm) wheelbase chassis that placed the side rails of the frame outside the seats, added to both safety and ride comfort.

Powering the Mark II was an overhead-valve V-8 that displaced 368 cu. in., at the time the biggest engine available in the United States except for the Packard 374. Top speed was documented at 117/118 mph (188/190 km/h). This impressive performance, matched with distinctive styling, demanded much in return—that is, from one's wallet. The Mark II cost approximately $10,000 when new, making it the most expensive motor car in America since the Duesenberg.

Opposite: Replete with stylish design details, the Continental Mark II was one of the most expensive cars in the world. At $10,000, it rivaled the Rolls-Royces of the day.

Right: The fully covered rear-mounted spare tire was carried over from the Mark II's predecessor. An effective image-builder for Lincoln, the Mark II was sold for just two model years, with approximately 3000 built in total.

1956 PONTIAC CHIEFTAIN COUPE

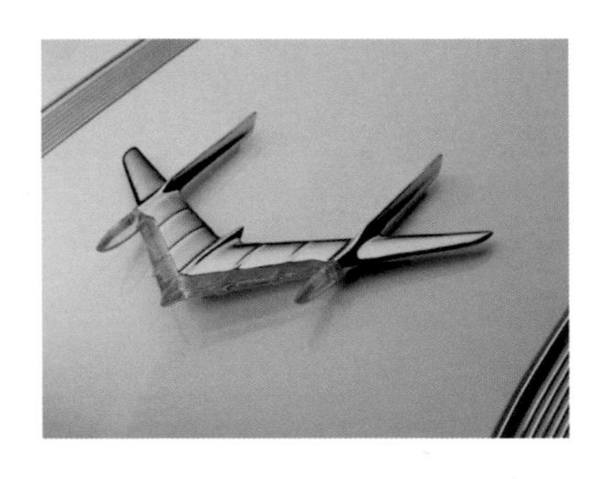

Pontiac's image was in desperate need of a makeover by the mid-1950s. A latecomer to the horsepower war raging all about Detroit, Dearborn, and South Bend, Pontiac finally declared in a press release of 1955 that it was the "ideal time" for entry into the V-8 field. The new Strato Streak V-8 was a fine piece of engineering, producing 180 hp and displacing 287 cu. in. while developing an admirable 264 lb ft of torque at 2400 rpm. Among its many benefits was the elimination of a conventional rocker-arm shaft and the need for manual valve adjustment via a ball-pivot valve train. Pontiac firsts for the engine included tapered valve guides, tin-plated pistons, and a harmonic balancer.

The new V-8 made the Chieftain a fairly fast family car. The 1955 Chieftains featured a completely new chassis and body, and a sales record of 546,531 units set that year, to which the Chieftain significantly contributed, was not surpassed until 1962.

For 1956, Pontiac added a controlled fluid coupling and more efficient clutches, calling the resultant transmission the Strato-Flight Hydra-Matic. The cars were now 2 3/8 in. (60 mm) longer—owing to a massive front bumper—while the tail fins were a little more pronounced. In addition, the capacity of the new-for-1955 V-8 was dramatically enlarged to 316 5/8 cu. in. Power increased considerably, jumping to 205 hp. The two Chieftain lines were renamed as the 860 and the 870, and a new body type—a four-door hardtop called the Catalina—was added to both lines.

Pontiac's budding performance image was given a shot in the arm in 1956 when speed ace Ab Jenkins drove a 285-hp Model 860 to a new NASCAR 24-hour distance record. Averaging 118.4 mph (190.5 km/h), Jenkins broke the previous record by 219 miles (352.4 km).

The Chieftain was one of the first all-new car designs to come from Pontiac in the postwar years.

PONTIAC

New styling for the 1956 Chieftain featured combination bumper grilles with enclosed circular parking lights and round, bomb-type bumper guards.

In addition to a slight redesign, the Chieftain of 1956 gained an increase in power. Beneath the substantial hood sat an enlarged V-8 producing 205 hp.

PONTIAC

1957 CHEVROLET BEL AIR CONVERTIBLE

At the beginning of the 1950s, Chevrolet was determined to hold on to its status as the top seller of domestic models, a position it had established before the United States had entered the Second World War. Its strategy was based on the Bel Air, its postwar "classy coupe" introduced in 1950. Sitting at the very top of the Chevy pecking order, the Bel Air was the division's trend-setting "pillarless" coupe, a fashionable "hardtop convertible" that set the pace for entry-level rivals Ford and Plymouth. Both competitors came out with respective copycat models in 1951.

It was the Bel Air that dominated the market niche, however. In 1953, the Bel Air line expanded to four models, with a true convertible and a pair of sedans joining the original two-door hardtop. The appeal of Bel Air ownership increased dramatically in 1955 after chief engineer Ed Cole and crew rolled out their vaunted "Hot One," the postwar classic powered by Chevy's new small-block overhead-valve V-8. The Bel Air moniker had hit its stride, selling more than 770,000 units in 1955.

By 1957, Chevrolet could say it had a class leader. That year's Bel Air lineup was characterized by crisp and clean styling inside and out. The exterior featured chrome headliner bands on hardtops, chrome spears on front fenders, chrome window moldings, full wheel covers, and the Bel Air script in gold lettering. Engine displacement grew to 283 cu. in., with the "Super Turbo Fire" V-8 option producing 283 hp thanks to continuous fuel injection.

Total Bel Air production for 1957 reached 702,220 units. Model choices had expanded to seven in 1956, following the addition of a four-door pillarless hardtop known as a "sport sedan." This seven-model lineup carried over into 1957, the last year in which the Bel Air reigned supreme atop the Chevy hierarchy. By that point, Chevrolet had built a total of 3.4 million Bel Airs.

The 1957 Chevrolet Bel Air is one of the most recognizable American cars of all time.

The Bel Air of 1957 increased in length by 2½ in. (64 mm). Optional extras included a two-tone interior, an automatic soft top, and ventilated seat pads.

475747
DELAWARE

1958 CHRYSLER 300-D

In the late 1950s, the man leading the charge for Chrysler was none other than automotive design legend Virgil Exner, a visionary whose talents were highlighted in the 300-D. The Chrysler 300 "letter series" cars were high-performance luxury automobiles built in limited numbers beginning in 1955. The 300 series achieved Chrysler's goal of reawakening interest in performance among major American manufacturers after the Second World War, doing so with an exclusive line that has since become recognized as one of America's finest postwar engineering and design efforts. The series moniker reflected the cars' performance: producing 300 hp, the 300s were the first American production cars to reach that milestone.

The Chrysler 300-D was the product of the fourth year of the marque's "letter series": high-performance luxury models built in limited numbers.

Under Exner's direction, Chrysler's competitive design focus consisted of new, upswept tail fins, a stylized grille, and a new chassis. Notably, in 1957, the 300-C sported for the first time its own exclusive front end, a definite departure from the everyday. Rectangular red-throated intakes beneath the headlamps channeled cooling air to the front brakes. Towering between them was a new fiberglass hood, fronted by a dominant trapezoid grille that was crammed edge to edge with an eggcrate design. These brazen elements carried over to the 1958 300-D.

Power was a big part of the 300s' identity. The 300-D was the last of the series to use the FirePower Hemi engine—392 cu. in. of muscle—although for 1958 it was tuned to 380 hp in standard form. Introduced in 1950 for the 1951 model year, the FirePower had hemispherical combustion chambers, essentially making it an early "Hemi" or "Generation 1 Hemi." Thirty-five 300-Ds were built with electronic fuel injection and delivered 390 hp. One 300-D reached 156.4 mph (251.7 km/h) at the Bonneville Salt Flats, which was good for publicity but did not result in a sales bump. In fact, only 618 hardtops and 191 convertibles were produced, thanks partly to 1958 being a recession year.

CHRYSLER

1959 CADILLAC

When Cadillac first introduced its tail fins (or "nubs") in 1948 (see page 155), it met with some resistance from consumers. But once customers realized their status value, complaints gave way to enthusiasm. As Cadillac moved into the 1950s, its styling became increasingly aggressive, with more pronounced features, including the flamboyant fins, the eggcrate grilles, and the vertical simulated air scoops on the rear fenders. Work on the models of 1959 began in the fall of 1956, with the car set for launch in October 1958.

The 1959 Cadillac was the third generation of the Eldorado. First seen in 1953, it became the longest-running personal-luxury-car series in America.

The 1959 Cadillac, which wielded the most outrageous fins ever seen on a production automobile, was something of a rite of passage for both the division and its parent, General Motors. That year saw the end of the Harley Earl era of design and the beginning of Bill Mitchell's tenure as design chief. The days of extravagance and unbridled optimism—vividly on display in the vast amounts of chrome and color on the models of the late 1950s—were transitioning to times of more restrained, "cleaner" models. Yet, for those who regarded the Dave Holls–designed Cadillacs of 1959 as over-the-top and ungainly, the design above and below the fins featured much cleaner elements, including airy greenhouses and considerably less chrome than in 1958. In fact, the 1959 models, which featured twin moon-rocket taillamps on each of the monstrous fighter-plane fins, outsold those of the year before: 142,000 units compared to 126,000.

The 1959 Caddies represented a garish symbol of success, and were as impressive on the road as they were controversial. Twin taillights were used because twice as many looked more expensive, and special nacelles were designed into the tail fins to accommodate them. In the front, the new grille was a treasure trove of toothy chrome, divided into upper and lower sections by a thin, horizontal blade. And the flow-through lines were indicative of what was on most designers' minds at the time: jet aircraft.

With its extravagant tail fins and bullet-shaped taillights, the 1959 Cadillac was instantly recognizable.

Opposite: Cadillac's tail fins reached their peak in 1959. An enduring symbol of America's unfettered optimism, they were seen by some as excessive.

Right: The industry's flamboyance of the late 1950s was well represented by the 1959 Cadillac, which featured new jewel-like grille patterns and matching rear panels.

BUICK
8728
Riviera

1963 BUICK RIVIERA

The big news from Buick in 1963 was its new luxury coupe, the Riviera—not to be confused with the body-type name that had been in use since 1949. This completely new model was an instant classic thanks to handsome and smart styling; its street presence was at once respectful and sporty. Interestingly, the Riviera originally was not intended to be a Buick, but rather a Cadillac, the "LaSalle II." The sporty elegance that resulted originated from the directive given to General Motors styling chief Bill Mitchell to create a "Ferrari-Rolls-Royce," a none-too-subtle crossbreed.

Touted as GM's "great new international classic car," the Buick Riviera took its styling cues from Rolls-Royce.

The car was unique among sister GM products of the early 1960s, and stood alone among American cars by providing just the right combination of luxury, road handling, and overall performance. It was a big Buick, yet its roadworthiness lent the driver an air of confidence at speed.

The Riviera was launched as a single model, a hardtop coupe. The Riviera's V-8 displaced 401 cu. in. and was rated at 325 hp at 4400 rpm. Sitting atop a 117-in. (2972-mm) wheelbase, the car rivaled Ford's Thunderbird of the day, prompting *Road & Track* to refer to the Riviera as "GM's T-Bird." For an additional $50, customers could order the optional bored-out version of the V-8, which was rated at 340 hp with a capacity of 425 cu. in.

Leather seats and a console came as standard. Exceptional styling was also standard. Unique frameless side-window glass accentuated a sharply sculptured roofline, and sheer side panels helped to create a low silhouette. Weighing in at just under 4000 lb (1814 kg), the Riviera came with a basic price of $4333.

Buick built 40,000 Rivieras in the car's first year of production, meeting all the firm's goals, including that of building prestige for the marque as it headed into the middle of the decade. Confirming this achievement was *Ward's Automotive Yearbook*, which noted that the Riviera was "a marvelously balanced prestige car."

Riviera

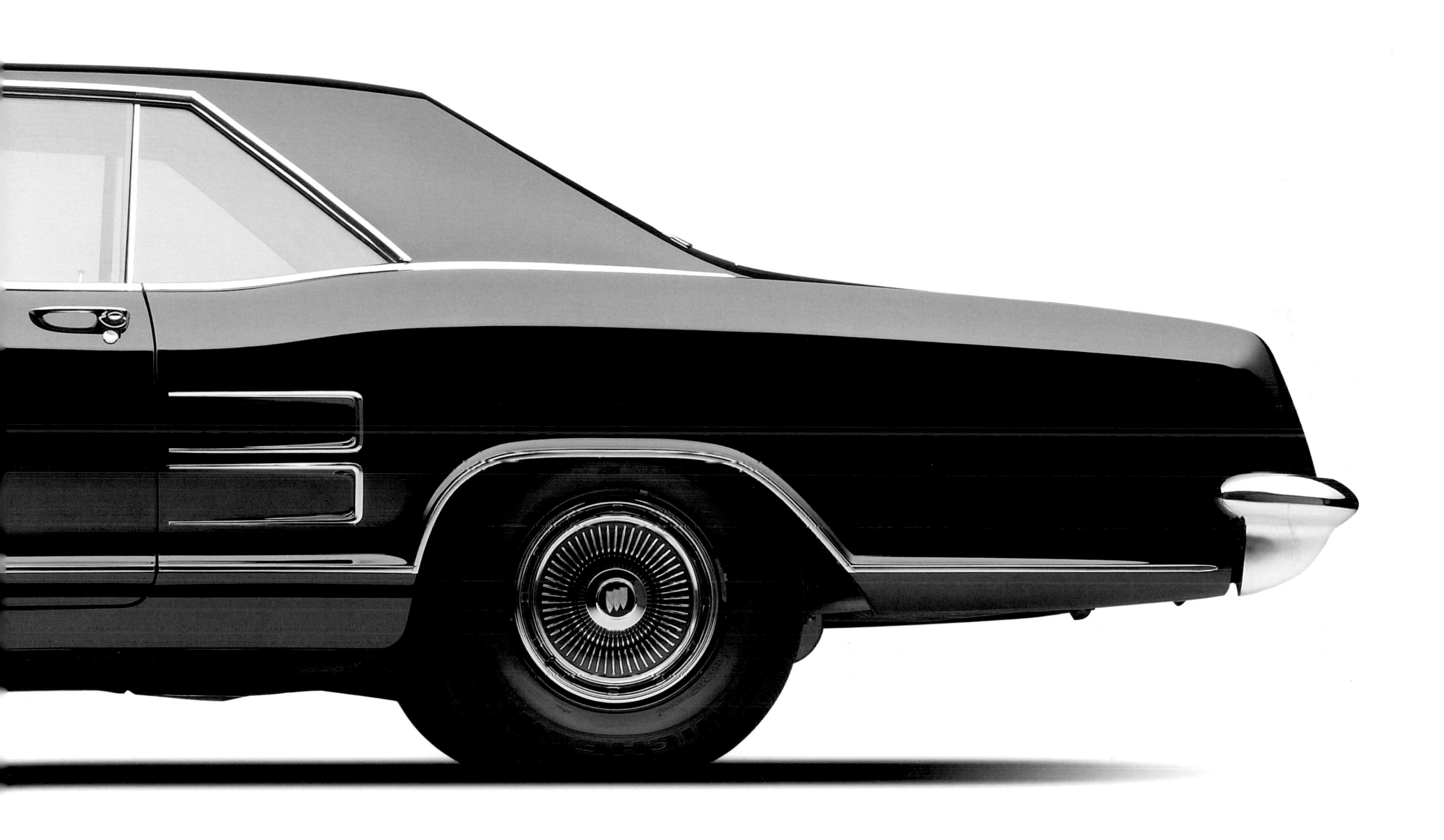

Left: Framed by a sharp, sculpted roofline, the Riviera's bold stance was accentuated by a low profile.

Opposite: Aggressive front-end styling was matched by a 325-hp V-8 powerplant under the hood.

1964 PONTIAC GTO

American automobiles are superb indicators of the country's mood, and this was especially true of the Pontiac GTO. For 1964, General Motors applied its B-O-P (Buick-Olds-Pontiac) approach used with its compacts to a new-sized car, the "intermediate." The company knew that the market was changing, and sought to beef up its compacts in size, power, and appointments. For Pontiac, this meant a second look at its Tempest/Le Mans range.

In addition to the ten models found in the expanded three Tempest lines of 1964 was a special model variant directed specifically at hardcore enthusiasts: the GTO. The previous year, 1963, had seen GM's board of directors issue a non-racing edict, which effectively ended Pontiac's official NASCAR participation, as well as its sanctioned drag racing. Pontiac's Pete Estes and John DeLorean, who understood that the firm's third-place market position was due in large part to its performance heritage, attempted to save the division's image. This involved fitting a high-performance, 389-cu.-in. V-8 into the lightweight, 3450-lb (1565-kg) Tempest.

The GTO was first offered as an option for the Le Mans. The basic package consisted of the 325-hp V-8 with a four-barrel carburetor and dual exhaust, a three-speed manual transmission with Hurst floor shifter, and a heavy-duty clutch. Stiffer coil springs and shock absorbers in the suspension gave the car a road-racer feel, as did a thicker front anti-roll bar. The Le Mans option package was a steal at $296. Additional costs were incurred for even racier options, including a 348-hp Tri-Power engine, a four-speed gearbox, metallic brake linings, and stiffer-yet suspension.

The efforts to keep Pontiac in the spotlight worked—and in a big way. Rival manufacturers rushed to produce their own asphalt eaters, and soon America found itself in the grip of the muscle-car revolution.

Pontiac's GTO resulted from the desire to stuff the largest available engine into the lightest possible body.

GTO

GTO
PONTIAC

The Pontiac GTO could complete the standing quarter-mile in just over 13 seconds, beating a stock 327 Corvette hands down. Once the GTO had become the talk of the town in Detroit, GM lifted its ban on big engines in intermediate-sized cars.

Left: Original production plans called for approximately 5000 Grand Turismo Omologatos to test the market. By early in the 1964 model year, however, sales figures had reached 10,000.

Opposite: Behind the 1964 GTO's unfussy grille sat either a 389-cu.-in., 325-hp engine with a single four-barrel carburetor, or the same engine with three two-barrel carburetors that raised the output to 348 hp.

GTO

BFGoodrich
Radial T/A
G.T. 350 H
BFGoodrich
Radial T/A

1966 FORD MUSTANG GT350H

Shelby American and Ford Motor Company joined forces to produce the GT350, Ford's vision for enhancing the Mustang's image as a performance car. Putting a street copy of the racecar on Hertz rental lots—and, by extension, on the streets in numbers—furthered the vision. Thus, the birth of the Mustang GT350H ("H" for "Hertz").

The "rent a racer" GT350H was a clever marketing ploy to place the high-performance Ford in the hands of Hertz customers.

Ford's proposal for a three-way cross-promotional advertising campaign including Ford, Shelby American, and Hertz was music to the ears of the rental company, which foresaw an onslaught of customers, most of them first-time renters, rushing to its counters. In December 1965, Shelby American received an order for 1000 GT350H models. The fulfillment of this order proceeded in fits and starts, however, with Hertz's requirements, particularly in terms of color schemes, becoming a moving target. The original order stipulated that the cars should be built as black fastbacks with gold rally stripes and gold rocker-panel stripes, but a later directive called for only the first 200 units to be finished in this way; the remaining 800 featured a mix of standard Shelby colors (white, red, blue, and green).

All Hertz cars were equipped with the GT350's normally aspirated 289-cu.-in. high-performance engine. Rated at 306 hp at 6000 rpm, and producing 329 lb ft of torque, this was the same engine that helped put Shelby at the top of the international racing heap. The first GT350Hs were equipped with four-speed manual transmissions; however, in an effort to tame the rough-and-tumble (and loud) driving experience, Hertz halted production of such models after delivery of the first eighty-five cars, and ordered automatics. Aside from the installation of a special brake booster, the GT350Hs were mechanically identical to non-Hertz GT350s.

With the GT350H, wannabe racers could have a slice of heaven for $17 a day and 17 cents per mile. Racing-class power in the guise of a rental car wildly fulfilled a once-in-a-lifetime marketing experiment.

BFGoodrich
Radial T/A
G.T. 350 H

BFGoodrich
Radial T/A

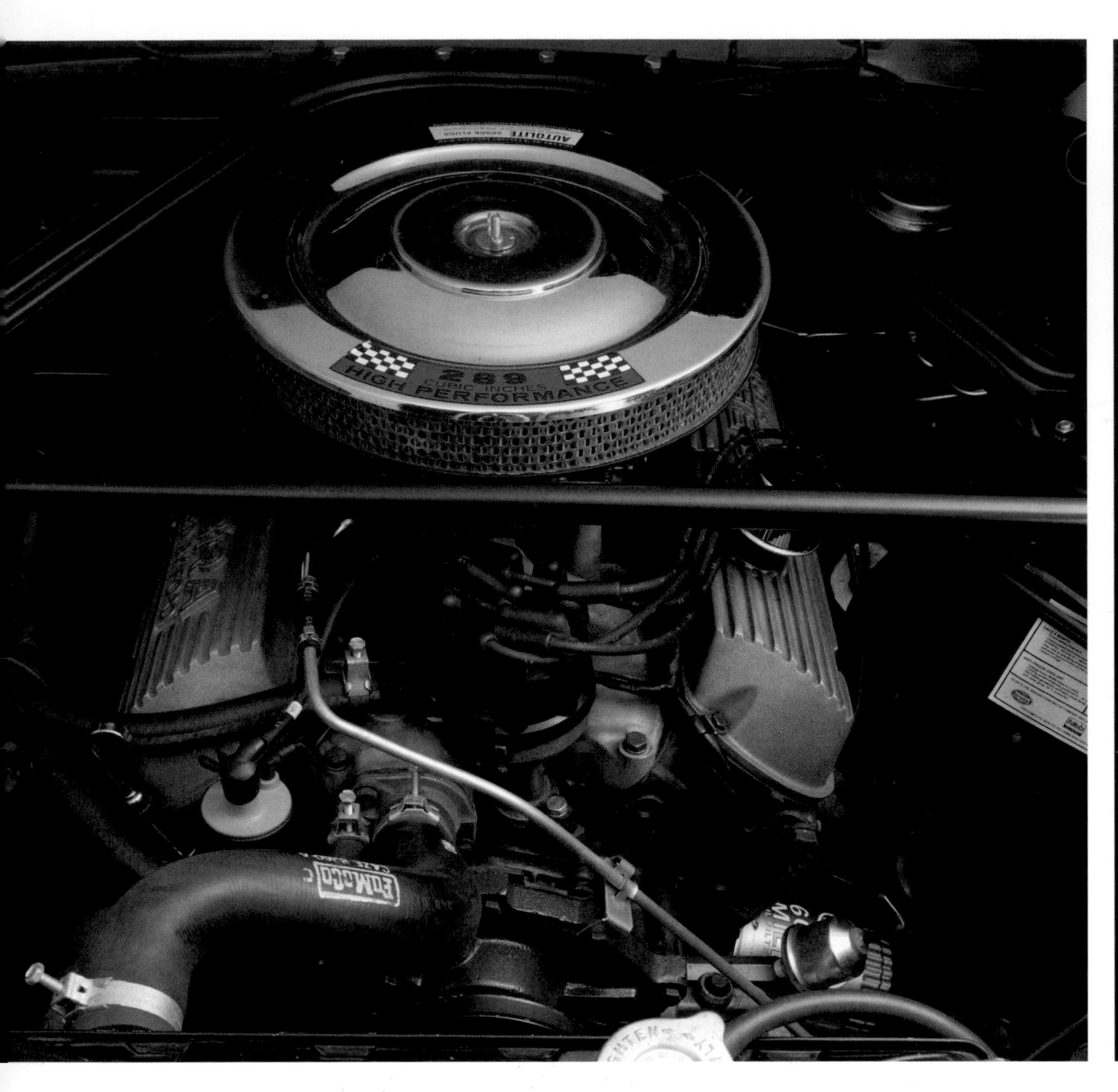

Above: Each GT350H was powered by a Cobra 289 high-performance V-8, a race-proven engine that produced 306 hp and 329 lb ft of torque.

Above, right: Shelby's Cobra logo was a welcome sign to enthusiasts looking for performance. With the GT350H, they were not disappointed.

Opposite: The Cobra-powered car was capable of 0–60 mph (97 km/h) in 6.6 seconds, and could complete the standing quarter-mile in 15.2 seconds at 93 mph (150 km/h). Its top speed was 117 mph (188 km/h).

Toronado

1966 OLDSMOBILE TORONADO

Something had to be done on Oldsmobile's part to compete with Buick's Riviera (see page 211), a sporty superstar from 1963 that garnered praise from buyers and the automotive press alike. Oldsmobile's answer was the front-wheel-drive Toronado, which debuted in 1966 and was awarded *Motor Trend*'s "Car of the Year" prize the same year. Based on General Motors' E-body platform, the well-engineered two-door coupe was conceived as Olds' full-size personal luxury car, as well as a direct competitor of Ford's popular Thunderbird (see page 185).

The Toronado's ace in the hole was its significance as the first front-wheel-drive automobile produced in the United States since the Cord of the 1930s. As opposed to the Cord's front-drive arrangement, which placed the engine slightly to the rear of the transaxle, the Toronado's driveline assembly—almost 1000 lb (454 kg) of it, including the engine—was positioned centrally over the front wheels. Creative engineering still allowed for a 54–46 front–rear weight distribution.

The Toronado's overall length was stretched to 211 in. (5359 mm), practically as long as a four-door sedan. Its appearance was strikingly singular, with a massive grille at the front and an exclusive roofline silhouette. Prominent side moldings focused the eye on the model's full-cut wheel arches. An aggressive overall front-end treatment impressed upon observers a notion of immediate power. And they were not disappointed. Olds used its 425-cu.-in. Rocket V-8 rated at 385 hp. Despite weighing in at almost 5000 lb (2268 kg), the 1966 Toronado could accelerate from 0–60 mph (97 km/h) in 7.5 seconds and complete the standing quarter-mile in 16.4 seconds with a terminal speed of 93 mph (150 km/h). Top speed was 135 mph (217 km/h).

The Toronado was GM's first "subframe" car, being only partly unitized. The subframe finished at the forward end of the rear-suspension leaf springs, serving as an attachment point for the springs. It also supported the powertrain, front suspension, and floorpan, resulting in better isolation of road and engine harshness.

The Toronado was ahead of its time, indicating the design revolution that would sweep the U.S. automotive industry in the 1980s.

Billed as a personal luxury coupe, the Toronado was a powerful car, as suggested by the muscular and aggressive front-end treatment.

The Toronado's prominent wheel arches are clearly visible from the rear. Oldsmobile spent seven years developing the car, during which time several GM innovations and designs were established.

GOODYEAR
G.T. 500
GOODYEAR

1967 SHELBY GT500

Ford's fiercest machine came in the form of the GT350 from Carroll Shelby's shop. But the car's racing attributes did not translate well to the street, its ride and noise more congruent with life on the track. Thus, Ford and Shelby collaborated on a refined version for the performance-minded, the Shelby GT500, which was introduced in November 1966 with a 428-cu.-in. V-8 with dual four-throat carburetors. The Cobra overhead-valve powerplant produced an advertised 355 hp, using a special high-revving hydraulic valve train and camshaft.

The 1967 GT500 featured a custom hood-and-grille assembly that recalled competition looks, none of which was simply decorative. Integral carburetor air scoops atop the hood increased the cooling area by 30 percent. Side treatments included sail-area air extractors that ventilated the driver's compartment, a feature inspired by Shelby's experiences at Le Mans. Custom lower-panel air scoops, a functional option seen on approximately the first 200 models, assisted with the cooling of the rear brake drums. The exclusive, sculptured grille and hood added 3 in. (76 mm) to the GT500's length. High-beam headlamps were mounted within the low, wide grille—initially in the center, and then at the sides—adding to the model's aggressive attitude.

Producing the GT500 had the desired effect, with Ford utilizing the Shelby name further to sell more Shelby Mustangs. In 1967, more GT500s (2050) were sold than GT350s (1175). That year's model set the precedent for future GT500s, although it was the last Mustang of the era to have Shelby's touch. In fact, the last few Shelby Mustang GT models of 1967 were made at A. O. Smith's facilities in Michigan—rather than Shelby American's shops in California—as production was moved closer to the Mustang's stomping grounds.

For Shelby, 1967 was an excellent year. Production was on the rise, with sales of the GT500 representing two-thirds of the extra demand. Carroll Shelby's increased need to assist Henry Ford II in Ford's racing efforts commanded more of his time, and the focus of in-house production on the GT500 proved very good timing.

The Shelby GT500 was developed as a road-friendly, high-performance alternative to Ford's Mustang.

COBRA
RADIO RESISTANCE

Opposite: The Shelby GT500 of 1967 came equipped with the 428-cu.-in. "Police Interceptor" power package.

Below: The GT500's design has endured for decades. The car's front-end treatment bears a striking resemblance to that of its modern-day descendant.

SHELBY
G.T. 500

Opposite: With Ford and Shelby always looking to improve performance, the GT500 was fitted with fiberglass body pieces, including an elongated nose.

Below: Clearly visible from the rear three-quarters are the body-side air intakes, the fiberglass tail with molded spoiler, and the sequential turn-signal lamps borrowed from the Mercury Cougar.

1969 CHEVROLET CORVETTE STINGRAY

The 1968 Sting Ray further set Chevrolet's Corvette apart, imprinting the undeniable brand on the minds of young and old yet again. A bold new design set the marque's destiny for years to come. The Stingray (now one word) of 1969 completed the transformation. That year marked the first time that an engine with a displacement of 350 cu. in. had been offered to Corvette. Rated at 300 hp, this standard unit was mated to a three-speed manual transmission. Popular optional power packages included the 427-cu.-in. L88, which produced 430 hp; four-speed transmissions with close-ratio gearing were also available.

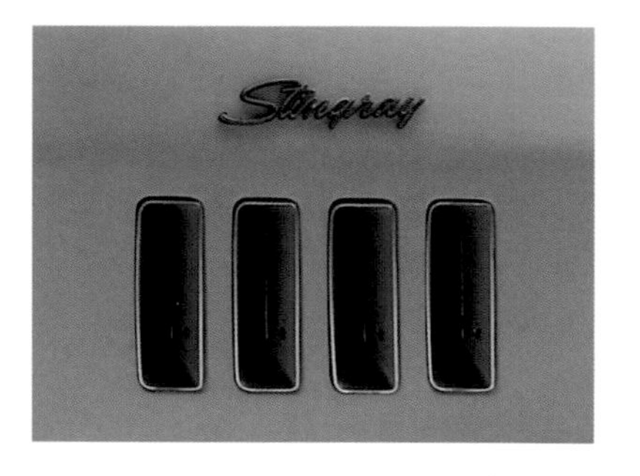

Stingray in both name and style, the 1969 Corvette was a personal favorite of General Motors design chief Bill Mitchell.

Further changes for 1969—in addition to a new "Stingray" badge on the front fenders—included an option for side-mounted exhausts and chrome trim on the side vents, as well as exterior door handles flush with the panel, as featured in the original design for the third-generation Corvette. As far as improvements over the previous year's model were concerned, other changes were subtle but substantial. One of the very few issues with the 1968 models was ride comfort, accentuated by shaking and shuddering at higher speeds, especially in the case of the open models. The suspension on the Stingray, a "luxury sports car," was optimally tuned for speeds in the range of 80–120 mph (129–93 km/h), and "Father of the Corvette" Zora Arkus-Duntov was unwilling to sacrifice such engineering for a plusher ride that would negate the brand's performance identity. For 1969, bumpy roads were smoothed somewhat with the help of rear-end frame stiffening. This was especially important, as the use of larger, wider wheels to maximize road adhesion and improve road handling made for a potentially harsher pounding.

Inside, the new Stingray delighted drivers with a cockpit feel. A neat tachometer was installed, and the console and panel reminded drivers and passengers alike that this was a sports car that emphasized spaciousness. And for the driver specifically, power steering was now available, underlining the "luxury" aspects of the Stingray. Standard-model production for 1969 totaled 22,129 sport coupes and 16,633 convertibles.

Opposite: In terms of looks, the Stingray was not shy in coming forward. Undulating side-panel moldings were both outrageous and sensuous.

Right: In addition to a stiffened frame for improved handling, the 1969 Stingray featured backup lights set within the inboard taillights.

Stingray

SS
Firestone
WIDE OVAL

1970 CHEVROLET CHEVELLE SS396

General Motors' approach to the muscle-car mayhem of the mid- to late 1960s was unique in that it took safety concerns very seriously. While Chrysler was packing monster 440-cu.-in. engines into its Mopar lineup and Ford was cramming a 428-cu.-in. powerplant into its small and mid-sized cars, GM imposed a 400-cu.-in. limit on all its compact and intermediate products. And although it shared the same platform as the Pontiac Tempest, the Chevelle Z-16, Chevrolet's answer to the GTO, did not appear until 1965. Powered by a version of the new 396-cu.-in. Mark IV big-block engine introduced in the Corvette and Impala that year, just 201 Z-16s were built. But the model served to whet the appetite of performance-hungry Chevy fans with a hint of things to come.

In 1966, the SS396 was made available as a standard production model, and instantly became the most popular mid-size performance machine of all, at least in terms of sales. Like those of the whole muscle-car market, sales of the SS396 peaked in 1969, when 86,307 were built. The car's legacy was high performance, macho styling, and good value. But in 1970, fans of the vehicle woke up to find a new offering: more power thanks to the lifting of GM's self-imposed 400-cu.-in. restriction. The SS396 handed over its crown as the bestselling muscle car on the planet to its successor, the LS-6 Chevelle, which used basically the same engine but with premium performance enhancements; the revised engine displaced 454 cu. in. and produced 450 hp. The LS-6 would be the swansong for Chevrolet's muscle cars.

Sales of the SS396 stayed strong throughout 1970, with 53,599 built. In fact, that year, it remained the most popular of the Chevelle performance cars.

Of all the iterations of the Chevrolet Chevelle—one of General Motors' most successful nameplates—the SS396 was one of the most popular.

Although it would prove to be one of the last of its kind, the Chevelle SS396 was all muscle car. To convince any doubters, it came packed with a 396-cu.-in. big-block V-8.

SS

1970 PLYMOUTH SUPERBIRD

For many manufacturers, proving oneself on the track was the precursor to success on the showroom sales floor. This was true for Plymouth and its Superbird as well. During its six years of production, Chrysler's street-version Hemi powerplant was offered as the ultimate performance option in most Dodge and Plymouth muscle cars. In competition, Chrysler had succeeded in qualifying the 426 Hemi as a production item, and was permitted to race it in the late 1960s. In 1970, the streamlined and spoilered Plymouth Superbird and Dodge Daytona both dominated circle-track racing, with Dodge taking the stock-car championship. However, the NASCAR sanctioning body felt that these "funny cars" had too much of an advantage, and for 1971 displacement was restricted to 305 cu. in. It was enough to make Chrysler withdraw from NASCAR competition.

Of the Superbird's many idiosyncratic features, a horn that sounded like the title character from the Road Runner cartoons was perhaps the most unusual.

The street-ready Superbird was produced in limited numbers for 1970. Essentially a highly modified Plymouth Road Runner, this larger-than-life muscle car featured a protruding, aerodynamic nosecone, a massive rear spoiler, and a horn that mimicked the Road Runner cartoon character.

The car's retractable headlights added 19 in. (483 mm) to the Road Runner's original length. Truly aerodynamic, the rear "wing" was mounted high enough to increase the efficiency of downdraft, although this design enhancement did not come into play unless speeds topped 90 mph (145 km/h). It is widely thought that the Superbird was produced in large part to lure racer Richard Petty back to Plymouth, after he had left the company's stock-car racing program for Ford in 1968.

With pollution regulations just around the corner and performance-car sales dying, 1970 would be the only year of the Superbird. It was available with one of three engine options: the 426 Hemi, the 440 Super Commando with a single four-barrel carburetor, or the 440 Super Commando Six Barrel with three two-barrel carburetors. Just under 2000 Superbirds were built.

Plymouth
BFGoodrich

The Superbird's bullet-style nose was based on that found on Dodge's Charger Daytona. All 'Birds were covered with a vinyl roof.

This most flamboyant of muscle cars was basically a highly modified Plymouth Road Runner. The most obvious modifications were the elongated nose, the fender scoops, and the enormous spoiler.

ROAD RUNNER
SUPERBIRD
Plymouth
1292

1971 PONTIAC TRANS AM

Pontiac was late to the pony-car party, finally making a bid to compete against the Ford Mustang and the Chevrolet Camaro with its Firebird of 1967. The Firebird was originally offered with either a six- or an eight-cylinder engine. The new look and performance of Pontiac's Firebird line for 1971 was too much for the General Motors division to keep bottled up, and the Firebird Trans Am was revealed as a $1970^{1}/_{2}$ model. On the same F-body platform as Chevrolet's Camaro, the Trans Am came with the Ram Air III engine fitted as standard; an ultimate 345-hp LS-1 round-port 400 was listed as an option. The top-of-the-line Firebird sported front and rear aerodynamic spoilers, rear-wheel "spats," front-fender vents, a throttle-controlled rear-facing "shaker" hood scoop, and many other special features.

Pontiac's bet on the Trans Am turned into a jackpot. In 1971, it surpassed the GM division's GTO as performance car of note.

Interestingly, the Trans Am was never intended for mass consumption. It was primarily intended to draw attention to the new Firebirds and thus generate showroom traffic; it was also a model designed to be raced, evidenced by Pontiac building just enough 3200 for the car to be homologated for SCCA Trans-Am series racing. Paying it the attention Pontiac so desired, *Car and Driver* noted that the Trans Am was "fantastically capable on a race track," while *Motorcade* described the car as "virile and aggressive in appearance."

The most significant changes to the Trans Am of 1971 were to be found under the hood. A mandate from GM called for engines to run on low-octane unleaded fuel in anticipation of forthcoming emission standards that would require catalytic converters. Meeting such a directive involved changing to low-compression ratios across the board, a serious blow to high-performance models. The Trans Am now came powered by a 455 HO engine, a high-end version of Pontiac's 455 complete with an aluminum intake manifold and many Ram Air IV components. Even with the new engineering, the Trans Am still generated a respectable 335 hp.

Equipped with the biggest V-8 ever offered in the pony-car field, the shapely yet solid Trans Am caused a stampede. More than 2100 Trans Ams were built for 1971, with a starting price of $4595. Unlike other muscle cars, the Trans Am proved resilient in the face of a shrinking market.

TRANS AM
GOODYEAR
POLYGLAS GT - F60-15

ROYAL SEAL
ROYAL SEAL

1985 CADILLAC SEVILLE

In the 1970s, Cadillac's most memorable creation was the Seville, a lower, slightly smaller Cadillac designed to appeal to younger buyers. With its 114-in. (2896-mm) wheelbase, the Seville proved that small did not necessarily mean cheap. When it was introduced in 1975, it was an immediate hit, and the style was copied by other domestic automakers. The Seville name had in fact first been used for a limited-production specialty model of the 1950s and 1960s.

The Seville was a pet project of Bill Mitchell, the vice-president for design at General Motors from 1959 to 1977. Mitchell had a youthful flair, and some of his other projects included the 1963 Buick Riviera (see page 211), the 1963 Corvette, and the 1970 Camaro. Over the years, Mitchell drew on many sources of inspiration, one of the most consistent being Rolls-Royce; he once said that the original Riviera was a combination of a Ferrari and a Rolls. "I'll never forget what my father once told me," he said in 1979. "'If you're going to steal . . . rob a bank, not a grocery store.'"

For the Seville of 1980—his last important design statement at GM—Mitchell directed the Cadillac designers to look at the Rolls-Royces of the 1930s and 1940s, with their sheer edges and slope-back styling. The result was the talk of the industry, a head-turner with its recumbent, "razor-edged" trunk. The Seville used the same front-wheel-drive chassis as the Eldorado (see page 166), the first front-wheel-drive Cadillac. Introduced at the Frankfurt Auto Show in October 1979, the 1980 Seville was the last of the flamboyant Cadillacs, part of an era that stretched back for decades.

Second-generation Sevilles—those produced between 1980 and 1985—retained the graceful but controversial bustle-backed body, and spawned such stylistic imitators as the Lincoln Continental Mark VI and the 1981 Imperial.

The Seville was an enduring nameplate for Cadillac. Its front-wheel-drive models of the 1980s helped the luxury marque weather a difficult economy.

1990 BUICK REATTA

By the time of its announcement in January 1985, Buick's plan for a two-seat luxury roadster was already making waves among the automotive press. The General Motors division had not had its own two-passenger car since before the Second World War, the 1940 Buick Model 46. Executives had been considering the potential market for a vehicle with the styling of a sports car and the comfort of a Riviera, and the result was the Reatta.

In January 1988, the Reatta coupe and Reatta convertible concept were unveiled at the Detroit Auto Show. The coupe was released to dealers later that month, introduced as a luxury two-seater that combined aerodynamic styling, agile handling, and a comfortable and quiet "Buick" ride. The powertrain offered at the time was the 3800 V-6 with four-speed automatic transmission. Incorporating elements of both a luxury car and a sports car, and priced at thousands of dollars below top-end versions of such vehicles, the Reatta was regarded by Buick executives as breaking new ground in a niche marketplace. Many items came as standard, including all-wheel power anti-lock brakes, Gran Touring suspension, and the Riviera's advanced CRT instrument panel with a touchscreen computer interface (a more conventional interface was fitted from 1990).

The Reatta had a unique body style, and was constructed with an emphasis on "hand finishing" via a process known as station assembly. In contrast to the common automobile built by workers as the vehicle passes along the assembly line, the Reatta was assembled as it moved from one station of workers to the next. To accentuate the departure from mass production, each model came with its own "Craftsman's Log" containing the signatures of the assembly supervisors. In 1990, a convertible was added to the Reatta lineup. Prices that year ranged from $28,235 to $34,995.

Buick's Reatta—offered as a convertible in 1990–91—was intended to be the marque's "halo" car. It boasted both performance styling and technical innovations.

1995 OLDSMOBILE AURORA

Oldsmobile was hurting in 1992, with sales having plummeted from 1,066,122 in 1985 to just 389,173. The General Motors division seemed to have lost sight of its customer, prompting it to produce a vehicle that was "not your father's Oldsmobile"—thus the introduction of the Aurora in 1995. This last gasp for survival served as a clean break from the other cars in the lineup: the Aurora bore no Oldsmobile badging or script, except on the cassette deck and engine cover. Instead, a new emblem consisting of a stylized "A" was used. Borrowing several styling cues from the Olds Toronado of the 1960s (see page 229), the Aurora was distinguished by full-width taillights, a wraparound rear windshield, and frameless windows.

The Aurora was positioned in the entry-level luxury class, and offered drivers a number of technologically advanced features, including dual-zone climate control and power-adjustable front seats with a two-position memory. Its standard powerplant was the 244-cu.-in. L47 V-8, a double-overhead-cam engine based on Cadillac's landmark 281-cu.-in. Northstar V-8, leading to high marks for the car's engine refinement. Performance was more than adequate, with 250 hp and 260 lb ft of torque. A well-balanced ride also scored points, but it was the excellent build quality and structural integrity that were most impressive. During normal crush-to-failure tests, conducted by automakers to evaluate body rigidity, the Aurora's unibody construction damaged GM's testing machine; a truck frame-crusher was used instead. The car exceeded federal safety standards for passenger cars by 200 percent.

First-year sales were strong, with the Aurora selling more than 45,000 units in 1995, but sales dropped dramatically the following year. For many car buyers, the Aurora was simply too expensive. More important, however, was the fact that the car did not "wow" the market to the point of reversing Oldsmobile's fate. Sales fluctuated, but the division finally succumbed in 2004.

The Oldsmobile Aurora breathed new life into the struggling General Motors division. Advertised as "not your father's Oldsmobile," the car took its styling cues from the 1960s Toronado.

2008 CADILLAC CTS

When Cadillac initiated its new "Art and Science" design direction, it chose to debut the new look on its just-contrived CTS. It proved a smart move, a perfect marriage of sharp styling with sharper-still engineering. Cadillac's production CTS benefited from the company's aggressive racing program of the 2000s. Winning the SCCA SPEED World Challenge Manufacturers' Championship was Team Cadillac's primary goal in 2005. The cars and drivers were up to the challenge: in what was only its second year on the track, the team claimed the championship after tallying four wins, two poles, and nineteen top-five finishes. At the Road Atlanta round of that year's competition, the CTS-*V* race team finished 1-2-3, besting rivals from the Porsche 911 and Dodge Viper outfits.

The release of the CTS is credited with reinvigorating the Cadillac brand. During the 1990s, Cadillac had attempted to capture a younger demographic with such models as the Catera and Allante, but the company did not achieve renewed success until the introduction of the CTS. With its stealth aircraft–inspired design, the car replaced the unsuccessful Catera.

In 2008, the CTS underwent a facelift. The new look was first seen at the 2007 Salon International de l'Auto in Geneva, when the 2008 CTS sedan made its European premiere. Its design inspired by Cadillac's Sixteen concept car, the new CTS reinterpreted the brand's traditional grandeur in a contemporary design. Advanced direct-injection gasoline technology helped a new 220-cu.-in. V-6 engine deliver 304 hp with improved fuel efficiency and reduced emissions.

"One of the coolest things we added is a subtle detail in the headlamps and taillights," explained CTS exterior designer Hoon Kim. "We used light pipes to create a contemporary vertical theme, front and rear. Really, it's show-car technology on a production car. But more important, it's a design cue that evokes classic Cadillacs like the Eldorado."

Bold and savvy, Cadillac's CTS was redesigned for 2008, maintaining the marque's new "Art and Science" design approach.

Reflecting Cadillac's racing-based approach of the early 2000s, the 2008 CTS came with rear-wheel drive as standard. All-wheel drive was made available for the first time.

2008 CHEVROLET CORVETTE

Introduced for the 2005 model year, the Chevrolet Corvette C6 platform is the latest in a series of products that can be traced back to Corvette's sports-car origins of 1953. For 2008, the C6 was fitted with a new powerplant, the LS3 V-8, which boasted such refinements as a larger, 378-cu.-in. capacity and a strengthened small block. When partnered with the optional low-restriction dual-mode exhaust system, the engine produced 436 hp—an increase of 36 hp compared to the previous iteration—and 428 lb ft of torque. A new high-flow cylinder head, larger valves, and an acoustically tuned intake manifold allowed drivers of the basic model to reach a top speed of 190 mph (306 km/h).

Inspired by the renewed horsepower race that had begun in the 1990s, Chevrolet decided to invest more time and effort in racing, applying advances made in the speed shops to its much-loved flagship sports car. The LS3 enabled the Corvette to do 0–60 mph (97 km/h) in 4.3 seconds, aided by quicker shift times on a returning sports paddle-shift for the six-speed automatic. Aesthetic refinements also abounded in the 2008 Corvette. Contributing to its first-glance identity were, at the front, dramatic fender forms and exposed headlamps that combined with the grille, and, at the back, a tapered deck. The lean rear design sported round taillights and a center-exit exhaust.

The Z51 performance package bestowed coupe and convertible models with performance levels that almost matched those of the previous generation Z06. The 2008 Z06 came with the new 427-cu.-in. LS7 engine, which delivered 505 hp in a 3132-lb (1421-kg) package, resulting in a 0–60 mph (97 km/h) time of 3.7 seconds in first gear, and a standing quarter-mile time of 11.7 seconds at 125 mph (201 km/h). With a top speed of 198 mph (319 km/h), the Z06 was the fastest and most powerful Corvette to date.

The 2008 Corvette was proof that Chevrolet's flagship model had not only survived but also thrived. The new car was fitted with a performance engine that could efficiently produce up to 436 hp.

Left: Refinements in driving characteristics were combined with the most aerodynamic body in the Corvette's history. Split-spoke cast-aluminum wheels came as standard; a forged-aluminum design was optional.

Opposite: The 2008 Corvette received several aesthetic modifications, including a tapered rear deck and fascia, a center-exit exhaust, and round taillights.

2008 PONTIAC SOLSTICE

If one had only products to consider, Pontiac would have seemed the least likely General Motors division to receive the ax in 2009. In the early and mid-2000s, Pontiac was billed as "Driving Excitement," and it was in that performance-based environment that the Solstice was born. In fact, it originated in the mind of GM vice-chairman Bob Lutz, the storied manager who envisioned a stylish, no-compromise, corner-gripping roadster that was affordable. The key word here was "affordable": the goal for the sticker price was $20,000.

In testing, the latest analytical tools were used to optimize performance characteristics, and the final product was approved as GM's best bet for competing head-to-head with the popular Mazda MX-5 Miata. The Solstice, an extremely light vehicle thanks to extensive use of aluminum throughout, was the first production car to use GM's efficient 146-cu.-in. LE5 Ecotech engine, the basic version producing 177 hp and 166 lb ft of torque. Tweaking the air-induction and engine-calibration components allowed the Solstice to beat the Miata's 0–60 mph (97 km/h) acceleration benchmark. The GXP version was fitted with a turbocharged 122-cu.-in. LNF Ecotech engine, which developed 260 hp and 260 lb ft of torque. The 2008 GXP was capable of 0–60 mph (97 km/h) in 5.5 seconds and 28 mpg (8.4 l/100 km), making the roadster even more affordable.

The Solstice concept car was introduced at the North American International Auto Show in Detroit in 2002, and audiences were quick to fall in love. It was the affirmation GM was looking for, and production planning began in earnest. Pontiac delivered 5445 models during the car's first year of production, 2005, and another 9700 in the first five months of 2006. Demand for the Solstice virtually outstripped plant capacity, and soon overtook that for its direct competitor, the Miata. The Solstice—along with the Saturn Sky, its sister make—set the stage for the successful, versatile Kappa platform, which features hydroformed side rails and a galvanized central tunnel.

With the closure of Pontiac in 2009, the rear-wheel-drive Solstice—in convertible and coupe form—proved to be one of the marque's final "Driving Excitement" cars.

You would be hard put to find an awkward angle on the Solstice—anywhere. Contributing to the near-organic lines are a long, European nose and properly sculpted flanks. The classic five-spoke alloy wheels further reinforce the car's sporty attitude.

A true driver's machine, the Solstice featured four-wheel independent suspension and GM's StabiliTrak electronic stability control system.

Chevrolet

DE LUXE CHAMPION

THE NAMES BEHIND THE LEGENDS

LOUIS CHEVROLET
1878–1941

Born in Switzerland and raised in France, Louis Chevrolet enjoyed racing bicycles as a young man, but lost his heart to the automobile at first sight. "What makes it run? How does it go? Can't it go faster?" he is said to have asked at the family dinner table. Later, a job with the De Dion Bouton Motorette Company took him to America, where he met William C. Durant (see below). Chevrolet and his brother Arthur became part of a successful team that designed and raced "Buick Bugs" for Durant and David Buick. Chevrolet then designed the first car to bear his name, the Chevrolet Classic Six, which was manufactured by Durant.

On November 8, 1911, Chevrolet partnered with Durant to start the Chevrolet Motor Car Company. After the two men fell out, Chevrolet formed the Frontenac Motor Company in partnership with Arthur and Gaston, his other brother. The firm's cars won many races, including the Indianapolis 500 in 1920 and 1921. Although the business was not a financial success, Chevrolet continued to make his expertise available to other manufacturers. He later turned to designing airplanes and aircraft engines.

WALTER P. CHRYSLER
1875–1940

From his first gainful employment—cleaning locomotives at 5 cents per hour—through his "clean-up" efforts at American Locomotive Company and Willys-Overland, Chrysler earned a reputation as a troubleshooter by analyzing problems and resolving them through decisive action. Building on his knowledge of shop practices gained as a laborer in the railroad industry, Chrysler easily spotted wasted material, time, and human resources. His workplace innovations, such as staggering shifts to improve efficiency and recycling engine oil for plant heating, turned his employers' losses into profits.

Chrysler's efficiency practices earned him the job of works manager for Buick Motor Company in 1912; by 1926, he had risen to the position of company president. In 1929, Chrysler became vice-president in charge of operations at General Motors, into which Buick had been integrated.

Chrysler's natural talents and a lifetime of learning paved the way for him to establish his own company at the age of fifty. Despite the late start, Chrysler succeeded where many pioneering independents had failed, building his firm into one of the auto industry's legendary "Big Three" (with GM and Ford).

FRED S. DUESENBERG
1876–1932

A German immigrant, Fred Duesenberg received no formal education in mechanics. As a youth, he and his brother, August, distinguished themselves by winning bicycle races. After learning about automobiles from automotive pioneer Thomas Jeffery, in Jeffery's Rambler factory, the Duesenberg brothers founded the Duesenberg Automobile and Motors Company in 1913, gaining international fame building powerful racing engines. Using his natural mechanical talents, Fred experimented with techniques that experienced engineers called "impossible." When he turned his attention to designing and building passenger cars, his imagination took flight. Although the Duesenberg brothers' enterprise suffered from financial problems, Fred's fabulous designs and engineering skill attracted financier E. L. Cord, who transformed the struggling business into a developer of classic cars that appealed to millionaires, movie stars, and royalty.

The slang expression "It's a doozie," used to compliment something extraordinary, is said to have derived from the nickname for Duesenberg cars.

WILLIAM C. DURANT
1861–1947

While traveling as a passenger in a sturdy two-wheeled cart, the young William "Billy" Durant inquired where it was built. On hearing the answer, he immediately boarded a train to the factory 75 miles (121 km) away, and later that day purchased the entire business for $1500. Already in debt, he convinced J. Dallas Dort to be his partner in the Durant-Dort Carriage Company.

In 1908, with little knowledge of automobiles, Durant established General Motors by consolidating dozens of small firms into a giant corporation. Forced out of the company in 1910 because of personal indebtedness, he bounced back, founding the Chevrolet Motor Car Company with Louis Chevrolet in 1911 (see above). He then bought more than 50 percent of GM's stock to regain control of the automaker. In 1920, having over-expanded the business, he was removed from power once again. Durant took only six weeks to get back into business, founding Durant Motors in January 1921.

CHARLES AND J. FRANK DURYEA
1861–1938 and 1869–1967

In 1896, the Duryea brothers founded the Duryea Motor Wagon Company, the first to manufacture and sell gasoline-powered vehicles in the United States. While other automotive pioneers were still tinkering with single vehicles, the Duryeas were making history. By the end of 1896, their company had produced thirteen gasoline-powered vehicles that were identical in design and construction, an accomplishment regarded as the beginning of the American automobile industry.

Charles Duryea was a visionary, predicting "the advent of the automobile" in his college thesis of 1882. Frank, the younger of the two brothers, was a talented mechanic. After brief careers in the bicycle business, both men collaborated on the development and production of a gasoline-powered vehicle. Their 1895 Motor Wagon won the now-legendary Chicago Times-Herald Race, completing the 54-mile (87-km) round trip from Chicago to Evanston, Illinois, in a time of 10 hours and 23 minutes. The car outperformed three German-made Benz models and two American electric cars. The publicity surrounding the Duryea victory inspired other mechanics and engineers to create and sell their own vehicles.

HARLEY J. EARL
1893–1969

Harley Earl, the son of a California custom-car builder, reshaped the look of automobiles from the ground up. Earl's career path became clear during a family camping trip in 1910. After a heavy rainfall, the sixteen-year-old Earl wandered from the campsite and found a clay-filled hollow. He then sculpted the clay into fantasy cars more elegant than any that existed at the time. Later, in the employ of his father, Earl designed opulent cars for Hollywood clients until a General Motors executive approached him with a job offer.

Following the success of the 1927 Cadillac LaSalle (see page 67), Earl was named manager of GM's revolutionary Art and Color Section, the first major styling department in the automotive industry. At GM, Earl introduced the use of clay models and obtained artistic freedom for himself and his staff in developing automotive designs. Under his guidance, GM stylists created vehicles that beautified the American roadway and permanently changed the shape of automotive product design.

VIRGIL M. EXNER
1909–1973

Virgil Exner lent a new shape to motion with innovative designs that made many American cars of the 1950s true "dream machines." Displaying artistic talent as a youth, Exner studied art at Notre Dame University, Indiana, and began his career in 1928 as an illustrator for Advertising Artists. He later joined General Motors' Art and Color Section, and at twenty-four was put in charge of the design of Pontiacs, becoming the youngest head of a GM styling division. In 1938, Exner joined Raymond Loewy's industrial design studio, where he was given the Studebaker account, designing some of the automaker's most popular models. But it was with Chrysler that Exner made his most lasting impression, developing the company's "Forward Look." Featuring streamlined bodies and tail fins, the "Forward Look" cars appeared ready for action even when parked, and were responsible for turning around Chrysler's slumping sales. Although Exner is often remembered as the "fin man," his talents extended to all areas of styling.

HENRY FORD
1863–1947

Born on a farm, Henry Ford hated the drudgery of agricultural work. On the road between Detroit and Dearborn one day, however, his life was changed forever: he saw a steam engine moving under its own power. Nearly twenty years later, he built a self-propelled vehicle of his own. His dream of building a car for the common man set Ford apart from other automotive pioneers. He founded the Ford Motor Company in 1903, and in 1908 the first of some 15 million Model T vehicles (see page 47) took to the road.

The Model T met the public's needs perfectly: inexpensive, reliable, easy to repair, and maneuverable on rough and muddy roads. Ford made the cars inexpensively and efficiently using an automated, moving assembly line. The repetitive work of the assembly line resulted in high employee turnover that reduced productivity. Ford responded when he announced the "$5-a-day workday"—about twice the going rate. Ford became a hero to workers who now could afford to buy their own car. In 1923, more than half of America's cars were Model Ts.

HENRY M. LELAND
1843–1932

Henry Leland brought style, grace, and a reputation for quality to the American automobile industry. Called the "Grand Old Man of Detroit," Leland entered the industry late in life after a successful manufacturing career. While working in the New England firearms factories of Colt and Brown & Sharpe, he had become acquainted with the concept of mass production using interchangeable parts. Leland's 1904 Cadillac was the first car to be built using such parts, promising owners an extended car life through more efficient repair and maintenance.

Because of Leland's insistence on precision manufacturing, the Cadillac name became synonymous with excellence and quality. Leland and his son, Wilfred, sold Cadillac to General Motors in 1909 but continued to operate the company until 1917, when they resigned from Cadillac in order to form the Lincoln Motor Company and create another legendary luxury vehicle.

RANSOM E. OLDS
1864–1950

Ransom Olds began his lifelong career building cars while still in his early twenties. In 1897, he founded Olds Motor Vehicle Company, which was reorganized in 1899 as Olds Motor Works. Olds was forced out of the company by partners who gained control and the use of the Olds name. By 1904, he was back in business, using his own initials to found the REO Motor Car Company.

The Curved Dash Oldsmobile, America's first mass-produced car, took to the streets in 1901. Readily available and highly functional, it proved to be a great success, gaining an endorsement for reliability when it became popular among doctors for visiting patients. Olds inspired a generation of young mechanics to explore the possibilities of the emerging auto industry. Although competitors infringed on many of his patents, Olds refused to prosecute them, and encouraged interested parties to visit his shop, observe his methods, and share his innovations.

WILLIAM DOUD AND JAMES WARD PACKARD
1861–1923 and 1863–1928

Before establishing their name in the automotive world, William and James Packard successfully operated the Packard Electric Company, which produced incandescent lightbulbs from its base in Warren, Ohio. Although James had made drawings of a "horseless carriage" as early as 1893, financial difficulties prevented the brothers from building their first automobile until 1899. In 1900, they applied for a patent for their innovative car design, which included a flexible shaft drive that could be used in place of the chain drive. The Packards formed the Ohio Automobile Company in 1900, changing the company's name to Packard Motor Car Company in 1902.

Always looking ahead, James and William Packard concentrated on securing resources for testing and design. Their company developed the first hook-up accelerator pedal, hand throttle, and automatic spark advance, as well as the first car to have three forward speeds and one reverse. The "Thirty-Eight" produced in 1914 introduced the grouping of lighting, ignition, and carburetor controls on the steering column, and featured left-hand drive and an electric starter.

CARROLL SHELBY
b. 1923

After a childhood marked by severe illness, Shelby's early career seemed to be heading nowhere. When, at the age of twenty-nine, his chicken-farming business failed, Shelby decided to do professionally what he had always enjoyed doing as a hobby: racing cars. He earned his first professional prize at Monza in 1954. The following year, he won ten U.S. races, then continued to win on both U.S. and European race circuits. Shelby was the first American to win the Le Mans 24 Hours, and was named sports-car "Driver of the Year" by *Sports Illustrated* in 1956 and 1957, and by the *New York Times* in 1957 and 1958.

Wearing his trademark bib overalls and participating in races even when injured, Shelby continued to enjoy success on the track until a heart condition forced him to retire. Turning his attention to design, he realized his dream of creating an affordable American sports car with the Shelby AC Cobra. Shelby's collaborations with Ford, and then Chrysler, brought high-performance cars within the budgets—and on to the driveways—of thousands of auto lovers.

ALFRED P. SLOAN
1875–1966

Alfred Sloan was responsible for transforming General Motors Corporation from a loose confederation of companies into an efficient, carefully coordinated, and successful manufacturer of motor vehicles. Selected as president of GM in 1923 and chairman of the board in 1937, Sloan served in the latter capacity until 1956.

After graduating from Massachusetts Institute of Technology, Sloan was hired as a draftsman at the Hyatt Roller Bearing Company, where he later became general manager. Recognizing that Hyatt could play a vital role in the emerging automotive industry, Sloan built and diversified the company over the next twenty years. When the company became part of GM, Sloan's talents were soon recognized. He was named vice-president and then, three years later, at a time when GM was still struggling to define itself, president. Sloan methodically examined all GM operations and molded the corporation into the largest automotive manufacturer in the world.

FRANCIS E. STANLEY
1849–1918

In 1897, Francis Stanley and his twin brother were operating a successful photographic supply business when F. E., as he was known, built a steam car for his own amusement. A new business opportunity developed the following year when a packed grandstand erupted in cheers as F. E. set a world speed record in the car, the Stanley Steamer. Suddenly, the brothers found themselves operating an unusual sideline to their photographic business, with customer orders for 200 vehicles.

In 1902, the Stanleys established the Stanley Motor Carriage Company, having sold the rights to the early version of their car to Locomobile. More than 10,000 Stanley Steamers, known for their quiet power, were built between 1897 and 1914. In 1906, one such Steamer reached 127 mph (204 km/h) to establish yet another speed record, one that stood for four years. The production of steam-powered cars eventually dwindled as gasoline-powered vehicles came into prominence.

ALEXANDER WINTON
1860–1932

Like those of so many auto pioneers of his day, Alexander Winton's roots were in the bicycle business. Winton, a Scottish immigrant, came to New York in 1878 at the age of nineteen, and worked for several years as an engineer on an oceangoing steamship. Soon tiring of the maritime life, he moved to Cleveland, Ohio, where he worked at an iron factory. In 1891, seeing a business opportunity, he founded the Winton Bicycle Company in Cleveland with the assistance of his brother-in-law. Despite the company's success, Winton was growing more interested in self-propelled vehicles.

In 1897, the Winton Motor Carriage Company was born. Although it built only four vehicles that year, the new firm established itself as one of the first automakers to be taken seriously. Of all the early automobiles, the Winton was considered to be the most technologically advanced—and the most powerful. A Winton was the first car to cross America coast to coast. Always the promoter, Alexander Winton entered as many races as he could, winning more often than not.

GLOSSARY OF MOTORING TERMS AND STYLES

A-pillar
In the side view, the foremost roof support of a vehicle, located in most instances between the outer edge of the windshield and the leading edge of the front-door upper. Also known as an A-post.

B-pillar
The roof support between a vehicle's front-door window and rear side window, if there is one. Also known as a B-post.

California top
A fixed rigid top applied to a touring car in place of the regular folding top, usually with sliding glass windows for weather protection.

camshaft
A shaft to which a cam is fastened, or of which a cam forms an integral part. In an internal-combustion engine, cams cause the cylinder valves to open as they rotate, either by pressing on the valves directly or via another mechanism.

chassis
A general term that refers to all the mechanical parts of a car attached to a structural frame. In cars with unitized construction, the chassis consists of everything but the body of the car.

coil spring
A type of spring made of wound heavy-gauge steel wire used to support the weight of a vehicle. The spring may be located between the control arm and chassis, the axle and chassis, or around a MacPherson strut. Coil springs may be conical or spiral, constant rate or variable rate, and wound with variable-pitch spacing or variable-thickness wire.

concealed headlamps
Headlamps that are hidden behind panels when not being used. When the headlamps are turned on, vacuum is applied to controllers that open the panels and expose the lamps. Also called "hideaway headlamps" or "pop-up headlamps."

concept car
A prototype designed to showcase new styling or technology. Some are merely paper drawings or clay mock-ups, while others are full-size vehicles for display at auto shows. While concept cars never go into production directly, some of the ideas they are designed to highlight may eventually appear on production models.

control arm
Part of a vehicle's suspension, a control arm is an integral component of a triangular "wishbone" configuration. The control arm pivots in two places, at the frame and at the steering knuckle.

coupe
Originally, a vehicle partitioned by a glass divider, fixed or movable, located behind the front seats. The driving position was only partially protected by the roof, while the totally enclosed rear was very luxurious. Modern coupes are two- to five-seaters with smaller interiors than sedans.

coupe de ville
A coupe with a completely open driving position.

C-pillar
The roof support between a vehicle's rearmost side window and its rear window. Also known as a C-post. On a vehicle with four side pillars, the rearmost roof support may be called a D-pillar, or D-post.

displacement
In an engine, the total volume of air/fuel mixture an engine is theoretically capable of drawing into all cylinders during one operating cycle. Generally expressed in cubic inches or liters.

double overhead camshaft (DOHC)
A DOHC engine has two camshafts in each cylinder head: one actuates the intake valves, while the other actuates the exhaust valves. The camshafts act directly on the valves, eliminating the need for pushrods and rocker arms.

dual cowl
A touring car with a folding cowl that covers part of the rear compartment and includes a rear windshield.

dynamometer
A machine used to measure the horsepower output of an engine. A chassis dyno has large rollers upon which a car's drive wheels are placed.

eggcrate grille
A radiator grille with crisscrossing bars that form square-like gaps. One of the distinctive features of Cadillacs.

fastback
A car with an unbroken curved line from the top of the roof to the rear bumper, as opposed to a line interrupted by a near-vertical rear window.

flagship
The prestige or top model of a manufacturer's line of vehicles, such as the Town Car for Lincoln and the CTS for Cadillac.

four-barrel carburetor
A carburetor with four barrels that work like dual carburetors, with the second carburetor (third and fourth barrels) cutting in only at high speeds. Usually found on large V-8 engines.

gear ratio
The number of revolutions a driving (pinion) gear requires to turn a driven (ring) gear through one complete revolution. For a pair of gears, the ratio is found by dividing the number of teeth on the driven gear by the number of teeth on the driving gear.

hand crank
Before Cadillac introduced the electric starter, cars were started by means of a handle that was inserted into the front of the engine and rotated manually. After 1930, the hand crank became obsolete.

hemi
Also called a "hemi head," an engine that uses hemispherical combustion chambers.

hot rod
A production car that has been modified by the owner in an attempt to increase acceleration and top-end speed. Such modification typically involves the use of a larger engine and the removal of some body panels. Although the term can be applied to any modified car, it is usually reserved for vehicles produced during the 1930s and 1940s.

independent suspension
A term used to refer to any type of suspension system that allows each of the two wheels of a given axle to move up and down independently of each other.

in-line engine
An engine in which all the cylinders (usually three or more) are arranged in a straight line (either vertically or slanted). The pistons drive a common crankshaft. Also called a "straight engine."

kick panel
A vertical interior panel enclosed by several structural members (for example, the side panel ahead of the A-pillar that extends up to the sides of the bulkhead and is limited by the floorpan at its bottom end).

knock-off
A single, large wing nut for fastening a wheel to its hub. Easily removed and replaced, it is "knocked off" by striking one of the wings with a mallet. Also called a "spinner."

lead-free gasoline
Gasoline with no lead additives, produced in response to concerns about the negative effects of lead on the environment and health, and to enable the use of catalytic converters. The additives, usually in the form of tetraethyl lead or some other lead compound, were originally introduced to increase the octane rating of the gasoline and reduce engine knock (detonation).

leaf spring
A simple form of suspension consisting of an arc-shaped length of flat spring steel, usually attached to the vehicle frame at both ends and to the axle in the center. Some applications need only one leaf spring, while most require several leaves, each smaller than the other, nested together.

L-head engine
An engine with both valves on one side of the cylinder.

louvers
A series of slits in the body of a vehicle that allows the air in and out. Also used on body panels for decoration.

magneto
A device that generates an electrical current when rotated by an external source of power. Once used as a source of ignition, it has since fallen out of favor in the automotive world.

mock-up
A full-size model of a vehicle made of wood and clay, used for design studies. *See also* concept car.

Mopar
The automobile parts (MOtor PARts) division of the Chrysler group. The name is also used more generally to refer to any Chrysler-built vehicle.

muscle car
Used to denote a high-performance car, the term is most often associated with American two-door, rear-wheel-drive mid-size vehicles of the late 1960s and early 1970s equipped with large, powerful engines and sold at affordable prices for street use and drag racing. Sometimes used interchangeably with "pony car" (see below).

phaeton
A body-style term taken from the era of the horse-drawn carriage. In the early days of motoring, it described a light car with large spoked wheels, with one double seat and, usually, a hood. The conventional phaeton—known as a double phaeton—has four doors, a convertible top, and side curtains rather than roll-up windows.

pony car
An affordable, compact, and styled vehicle—characterized by a sporty or performance-oriented image—inspired by the Ford Mustang of mid-1964. *See also* muscle car.

pushrod
Part of an overhead-valve engine, pushrods trigger rocker arms above the cylinder heads to actuate the valves.

roadster
A two-seater with an open body and a folding fabric top. Other defining features can include a luggage compartment or rumble seat at the rear.

rocker arm
Within an internal combustion engine, rocker arms are reciprocating levers used to actuate the cylinder valves. They are operated by the radial movement of the camshaft, either directly or via pushrods.

rpm
An abbreviation of 'revolutions per minute', rpm is used to measure the rotational speed of a mechanical component. In an automotive context, it usually refers to the crankshaft.

runabout
A small, open, sporty type of vehicle, generally with only two seats and simple bodywork.

sedan
A closed body type with four doors (but sometimes two, or possibly three or five in the case of a hatchback), four or six windows, and seating for four or more passengers.

skiff
An open sports car with streamlined, light bodywork.

spoiler
An aerodynamic device, normally on the rear of the vehicle, that disrupts ("spoils") the flow of air over the vehicle's body in order to reduce drag and increase fuel efficiency. Sometimes confused with a wing, the main purpose of which is to create downforce and improve road handling.

sports car
An agile vehicle that is easily maneuverable, accelerates briskly, brakes positively, and steers precisely. It is tightly sprung and usually does not offer a comfortable ride.

tonneau
An open vehicle with a bench seat in the front and a semicircular seat behind. On early examples, access to the rear seat was through a rear-facing door built into the seat itself.

torque
Generated by the engine and measured in pounds per feet (lb ft), torque is the twisting force, or "push," that sets a vehicle in motion and increases its velocity. Specification charts usually include the maximum torque the engine can develop, and the rpm at which it is generated.

touring car
An open car with seats for four or more passengers. Early models had no weather protection at the sides, but were later fitted with detachable side screens and curtains.

underslung
A term used to describe the suspension of a vehicle's frame below its axles.

unibody construction
A type of body construction that does not require a separate frame to provide structural strength or support for the vehicle's mechanical components.

Victoria
A body style derived from the era of horse-drawn carriages. A Victoria was long and luxurious, with a separate driving position and a large rear seat, and was equipped with hoods and side screens.

wheelbase
The distance between the centers of the front and rear axles. Wheelbase is important because it indicates available body length and weight distribution between both axles.

Woodie (or Woody)
A vehicle with a partly wooden frame, usually a station wagon of the 1930s or 1940s. In the 1960s, woodies became popular among the surfing community.

DIRECTORY OF MUSEUMS AND COLLECTIONS

Antique Automobile Club of America Museum
161 Museum Drive
Hershey, PA 17033
Tel: +1 717 566 7100
aacamuseum.org

Auburn Cord Duesenberg Automobile Museum
1600 South Wayne Street
Auburn, IN 46706
Tel: +1 260 925 1444
automobilemuseum.org

Automotive Hall of Fame
21400 Oakwood Boulevard
Dearborn, MI 48124
Tel: +1 313 240 4000
automotivehalloffame.org

Blackhawk Museum
3700 Blackhawk Plaza Circle
Danville, CA 94506
Tel: +1 925 736 2277
blackhawkmuseum.org

Classic Car Club of America Museum
P.O. Box 2113
Dearborn, MI 48123
Tel: +1 269 353 4672
cccamuseum.org

Crawford Auto Aviation Museum
10825 East Boulevard
Cleveland, OH 44106
Tel: +1 216 721 5722
wrhs.org

Gateway Colorado Automobile Museum
43224 Highway 141
Gateway, CO 81522
Tel: +1 970 931 2895
gatewayautomuseum.com

Gilmore Car Museum
6865 Hickory Road
Hickory Corners, MI 49060
Tel. +1 269 671 5089
gilmorecarmuseum.org

GM Heritage Center
6400 Center Drive
Sterling Heights, MI 48312
Tel: +1 586 276 0695
gm.com

Indianapolis Motor Speedway Hall of Fame Museum
4790 West 16th Street
Indianapolis, IN 46222
Tel: +1 317 492 6784
indianapolismotorspeedway.com

Lane Motor Museum
702 Murfreesboro Pike
Nashville, TN 37210
Tel: +1 615 742 7445
lanemotormuseum.org

LeMay Museum
325 152nd Street East
Tacoma, WA 98445
Tel: +1 253 536 2885
lemaymuseum.org

National Automobile Museum
The Harrah Collection
10 South Lake Street
Reno, NV 89501-1558
Tel: +1 775 333 9300
automuseum.org

The National Corvette Museum
350 Corvette Drive
Bowling Green, KY 42101
Tel: +1 270 781 7973
corvettemuseum.com

National Packard Museum
1899 Mahoning Avenue NW
Warren, OH 44483
Tel: +1 330 394 1899
packardmuseum.org

The Nethercutt Museum
15151 Bledsoe Street
Sylmar, CA 91342
Tel: +1 818 364 6464
nethercuttcollection.org

Owls Head Transportation Museum
117 Museum Street
Owls Head, ME 04854
Tel: +1 207 594 4418
ohtm.org

Petersen Automotive Museum
6060 Wilshire Boulevard
Los Angeles, CA 90036
Tel: +1 323 930 2277
petersen.org

Simeone Foundation Museum
6825-31 Norwitch Drive
Philadelphia, PA 19153
Tel: +1 215 365 7233
simeonefoundation.org

Studebaker National Museum
201 South Chapin Street
South Bend, IN 46601
Tel: +1 574 235 9714
studebakermuseum.org

Walter P. Chrysler Museum
1 Chrysler Drive
Auburn Hills, MI 48326-2778
Tel: +1 248 944 0001
chryslerheritage.com

INDEX

Main entries for cars, designers, and automobile-company founders are shown in **bold**.